T0216943

SpringerBriefs in Complexity

SpringerBriefs in Complexity are a series of slim high-quality publications encompassing the entire spectrum of complex systems science and technology. Featuring compact volumes of 50 to 125 pages (approximately 20,000–45,000), Briefs are shorter than a conventional book but longer than a journal article. Thus Briefs serve as timely, concise tools for students, researchers, and professionals.

Typical texts for publication might include:

- A snapshot review of the current state of a hot or emerging field
- A concise introduction to core concepts that students must understand in order to make independent contributions
- An extended research report giving more details and discussion than is possible in a conventional journal article,
- A manual describing underlying principles and best practices for an experimental or computational technique
- An essay exploring new ideas broader topics such as science and society

Briefs allow authors to present their ideas and readers to absorb them with minimal time investment. Briefs are published as part of Springer's eBook collection, with millions of users worldwide. In addition, Briefs are available, just like books, for individual print and electronic purchase. Briefs are characterized by fast, global electronic dissemination, straightforward publishing agreements, easy-to-use manuscript preparation and formatting guidelines, and expedited production schedules. We aim for publication 8–12 weeks after acceptance.

SpringerBriefs in Complexity are an integral part of the Springer Complexity publishing program. Proposals should be sent to the responsible Springer editors or to a member of the Springer Complexity editorial and program advisory board (springer.com/complexity).

Santo Banerjee · A. Gowrisankar ·
Komandla Mahipal Reddy

Fractal Patterns
with MATLAB

 Springer

Santo Banerjee
Dipartimento di Scienze Matematiche
Politecnico di Torino
Turin, Italy

A. Gowrisankar
Department of Mathematics, School
of Advanced Sciences
Vellore Institute of Technology
Vellore, Tamil Nadu, India

Komandla Mahipal Reddy
Department of Mathematics
VIT-AP University
Amaravati, Andhra Pradesh, India

ISSN 2191-5326   ISSN 2191-5334 (electronic)
SpringerBriefs in Complexity
ISBN 978-3-031-48101-7   ISBN 978-3-031-48102-4 (eBook)
https://doi.org/10.1007/978-3-031-48102-4

This Springer imprint is published by the registered company Springer Nature Switzerland AG
The registered company address is: Gewerbestrasse 11, 6330 Cham, Switzerland

Paper in this product is recyclable.

Preface

Nature exhibits "formless" patterns with different levels of complexity, for instance, imagine clouds, coastlines and mountains which are not spheres, circles and cones as Mandelbrot quoted in the book *The Fractal Geometry of Nature*. He coined the word *fractal* to describe challenging irregular and fragmented patterns. As fractal portrays natural phenomena, its elegance lies in its colorful graphics and intricate patterns. When fractals are browsed on the internet, we get colorful illustrations of never-ending patterns preserving self-similarity across any small scale. Beginning with mobile wallpapers, fractals can be seen in every turn of nature including medical images, human anatomy and river networks. The field of *Fractal Geometry* benefits the readers with fascinating visuals of fractal sets and fractal functions rather than providing rigorous theoretical background.

This book is a visual treat for fractalists as well as for non-fractalists with elementary MATLAB coding. Patterns of deterministic fractals, fractal functions and fractal surfaces are illustrated using MATLAB. Chapter 1 briefly discusses the construction of a deterministic fractal and its iteration algorithm. The examples of deterministic fractals such as Sierpinski triangle, von Koch curve and so on are described mathematically and their patterns are exemplified with MATLAB code. Further, MATLAB code for estimating fractal dimensions is presented for box counting, Higuchi and Katz fractal dimensions.

Chapter 2 presents graphical illustrations of univariate fractal interpolation functions with constant and variable scaling factors. The MATLAB code is provided for generating fractal curves at different levels of iterations with the given data set and scaling parameters. Chapter 3 benefits the readers with appealing fractal graphs of differentiable fractal interpolation functions. For the prescribed shape parameters and derivative values, MATLAB code is prescribed for achieving different fractal splines.

Chapter 4 precisely recalls the construction of fractal surfaces. MATLAB code is presented for generating a variety of fractal surfaces along X-axis and Y-axis depending on the shape and scaling parameters. Chapter 5 explores the application of fractal functions. Graphs of mountains and clouds are approximated using MATLAB code with different sets of scalings. In addition, the first and second waves of Omicron are reconstructed using affine fractal interpolation functions.

Torino, Italy   Santo Banerjee
Vellore, India   A. Gowrisankar
Amaravati, India   Komandla Mahipal Reddy

Contents

List of Figures

List of Tables

Chapter 1
Fractals and Dimensions

1.1 Introduction

In Euclidean geometry, any smooth curve is one-dimensional and any smooth surface is two-dimensional. It is not the case when we encounter with natural objects, nature is completely spread with its own randomness everywhere. There are curves and surfaces with high irregularities that cannot be investigated using the Euclidean geometry. The birth of fractal geometry is a boon to analyze such non-smooth natural curves and surfaces. Mandelbrot, who introduced the word *fractal* has mathematically defined it *as a set with the Hausdorff dimension strictly exceeds its topological dimension*, refer [1].

An *iterated function system* (IFS) is the fundamental concept behind the construction of fractals, it generates the deterministic fractal as its unique attractor. An IFS is a family of finite number of continuous maps on a complete metric space and it is hyperbolic if all the continuous maps are contractive. Let (X, d) be a complete metric space. The map $w : X \to X$ is said to be a contraction map if it satisfies

$$d(w(u), w(v)) \leq cd(u, v), \text{ for all } u, v \in X,$$

where c is called the contraction ratio or contractivity factor such that $c \in [0, 1)$. A point $a \in X$ is called the fixed point of w if $w(a) = a$. In general, a contraction map may possesses any number of fixed points and it need not be unique. However, the Banach fixed point theorem guarantees the existence of a unique fixed point when contractions are defined on a complete metric space. Consider a complete metric space X with respect to the metric d. Let $\mathcal{H}(X)$ be the set of all non-empty compact subsets of X. The Hausdorff metric h on $\mathcal{H}(X)$ is defined by

$$h(A, B) = \max\{\sup_{u \in A} \inf_{v \in B} d(u, v), \sup_{v \in B} \inf_{u \in A} d(u, v)\},$$

© The Author(s), under exclusive license to Springer Nature Switzerland AG 2023
S. Banerjee et al., *Fractal Patterns with MATLAB*,
SpringerBriefs in Complexity,
https://doi.org/10.1007/978-3-031-48102-4_1

for $A, B \in \mathcal{H}(X)$. The hyperspace $(\mathcal{H}(X), h)$ is a complete metric space, provided that (X, d) is complete. For $k = 1, 2, \ldots, N$, defining the continuous maps w_k as self-maps on $\mathcal{H}(X)$, the system

$$\{X; w_k : k = 1, 2, \ldots, N\} \tag{1.1}$$

constitutes an IFS. Combining all the contraction maps w_k defined on $(\mathcal{H}(X), h)$, a new contraction map is produced, namely Hutchinson-Barnsley operator $W : \mathcal{H}(X) \to \mathcal{H}(X)$. It is defined by

$$W(B) = \bigcup_{k=1}^{N} w_k(B),$$

where $B \in \mathcal{H}(X)$ and $w_k(B) = \{w_k(u) : u \in B\}$. If the continuous maps w_k obey

$$d(w_k(u), w_k(v)) \leq r_k d(u, v), \text{ for all } u, v \in X,$$

where r_k are contractivity factors such that $0 \leq r_k < 1$, then the IFS (1.1) is referred as the hyperbolic IFS in [2]. It follows that W is also a contraction mapping with respect to the Hausdorff metric satisfying

$$h(W(A), W(B)) \leq r h(A, B), \text{ for all } A, B \in \mathcal{H}(X),$$

where $r = \max\{r_k : k = 1, 2, \ldots, N\}$ is the contractivity factor. Then W possess a unique fixed point G such that

$$G = W(G) = \bigcup_{k=1}^{N} w_k(G).$$

In addition, the unique fixed point $G \in \mathcal{H}(X)$ obeys

$$G = \lim_{n \to \infty} W^{\circ n}(A), \text{ for each } A \in \mathcal{H}(X),$$

and it is called as the *attractor* or *deterministic fractal*, where $W^{\circ n} = W \circ W \circ \cdots \circ W$(n times) is the n-fold auto-composition of the map W. For more details on the construction of deterministic fractal, refer [2, 3].

1.2 Deterministic Iteration Algorithm

Barnsley has proposed two algorithms namely the deterministic algorithm and the random iteration algorithm for computing fractals in [2]. In this section, deterministic iteration algorithm is briefly recalled. Consider the *IFS* $\{X; w_k : k = 1, 2, \ldots, N\}$

with the Hutchinson-Barnsley operator W on $\mathcal{H}(X)$. Choose an initial non-empty compact set of X, say G_0 such that the sets are iteratively computed as follows

$$G_1 = W(G_0) = \bigcup_{k=1}^{N} w_k(G_0)$$

$$G_2 = W(G_1) = \bigcup_{k=1}^{N} w_k(G_1)$$

$$\vdots$$

$$G_n = W(G_{n-1}) = \bigcup_{k=1}^{N} w_k(G_{n-1}).$$

It is well-known that the sequence of sets $G_0, G_1, \ldots, G_n, \ldots$ converges to a non-empty compact set G (attractor of the *IFS*) by the Banach fixed point theorem.

For any choice of initial non-empty compact set in X, the algorithm yields the attractor G. However, if G_0 is completely unrelated to the attractor, the number of iterations will be larger for the convergence.

1.2.1 Sierpinski Triangle

The exact self-similarity of the Sierpinski triangle at any small scale makes it an interesting and simple example of classic deterministic fractals on two-dimensional space. Using the iterative approach, the Sierpinski triangle is constructed as follows. Consider the larger equilateral triangle of side length l, joining the midpoints of each side, one can obtain four new equilateral triangles with side length being equal to $l/2$, delete the centre triangle leaving the boundary. As a result of first iteration, three equilateral triangles are obtained. In the second iteration, the midpoints of three sides of each small triangle are joined to remove the centre triangle from each of the three, thus nine equilateral triangles are obtained, each of side length $l/4$. The same process is applied recursively with the remaining smaller triangles and at the nth iteration, 3^n number of triangles are obtained with side length $(l/2)^n$. The infinite intersection of all the equilateral triangles obtained in each iteration constitutes the Sierpinski triangle. The entire length of the Sierpinski triangle is found to be infinite since its total length $(3/2)^n$ is calculated as a series diverging to infinity i.e., $\lim_{n \to \infty}(3/2)^n$. For detailed description, refer the books [4, 5].

To construct a Sierpinski triangle through the concept of IFS, consider the contractive mappings w_1, w_2 and w_3 with contractivity factor $1/2$ defined on $[0, 1] \times [0, 1]$ as given below,

$$w_1 \begin{pmatrix} x \\ y \end{pmatrix} = \begin{pmatrix} 0.5 & 0 \\ 0 & 0.5 \end{pmatrix} \begin{pmatrix} x \\ y \end{pmatrix} + \begin{pmatrix} 0 \\ 0 \end{pmatrix},$$

$$w_2 \begin{pmatrix} x \\ y \end{pmatrix} = \begin{pmatrix} 0.5 & 0 \\ 0 & 0.5 \end{pmatrix} \begin{pmatrix} x \\ y \end{pmatrix} + \begin{pmatrix} 0.5 \\ 0 \end{pmatrix}, \qquad (1.2)$$

$$w_3 \begin{pmatrix} x \\ y \end{pmatrix} = \begin{pmatrix} 0.5 & 0 \\ 0 & 0.5 \end{pmatrix} \begin{pmatrix} x \\ y \end{pmatrix} + \begin{pmatrix} 0 \\ 0.5 \end{pmatrix}.$$

Hence, $\{[0, 1] \times [0, 1]; w_i : i = 1, 2, 3\}$ constitutes the *IFS* whose attractor is the required Sierpinski triangle. The following is the MATLAB code for generating the Sierpinski triangle using contraction maps (1.2) in the aforementioned deterministic iteration algorithm (Fig. 1.1).

```
% The Sierpinski triangle (or Sierpinski gasket)
%W=AX+B
%where A=[0.5 0;0 0.5];X=[x;y];B=[0 0.5 0;0 0 0.5];
% initial X is [0 1 0;0 0 1];
clc;clear all;close all;
X=[0 1 0;0 0 1];A=[0.5 0;0 0.5];B=[0 0.5 0;0 0 0.5];% input
Termination=6;
for iter=1:Termination
    if iter==1
        w=A*X+B;
        X=w;
        XX=[X X(:,1)];
        x=XX(1,:);y=XX(2,:);
        plot(x,y)
        fill(x,y,'b')
    else % more than one iteration figure
        w1=A*X+B(:,1);%First tranformation
        T1=w1;
        w2=A*X+B(:,2);%Second transformation
        T2=w2;
        w3=A*X+B(:,3);%Third transformation
        T3=w3;
        X=[T1 T2 T3];
        [m n]=size(X);
        for i=1:n/3
            D=[X(:, 3*(i-1)+1:3*i) X(:,3*(i-1)+1)];
            x=D(1,:);y=D(2,:);
            plot(x,y)
            fill(x,y,'b')
            hold on
        end
    end
end
```

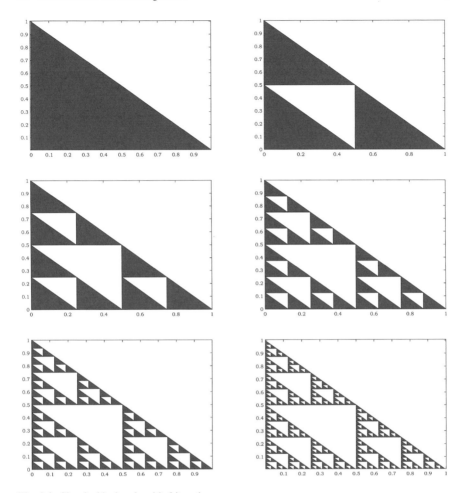

Fig. 1.1 Sierpinski triangle with 6 iterations

1.2.2 von Koch Curve

Let the unit interval [0, 1] be denoted by J and a line segment of unit length by J_0. The construction of von Koch curve begins with the removal of middle third segment of J and replacing it by other two sides of the equilateral triangle based on the removed segment. Set J_1 is obtained. Next, set J_2 is generated by applying the similar procedure to each segment of J_1 and so on. The limiting set is called von Koch curve. We can model this construction using an iterated function system consisting of contractions $\{w_1, w_2, w_3, w_4\}$ on the the square $[0, 1] \times [0, 1]$, where transformations are given by

$$w_1(x, y) = \left(\frac{x}{3}, \frac{y}{3}\right),$$

$$w_2(x, y) = \left(\frac{2 + x - \sqrt{3}y}{6}, \frac{\sqrt{3}x + y}{6}\right),$$

$$w_3(x, y) = \left(\frac{3 + x + \sqrt{3}y}{6}, \frac{-\sqrt{3}x + y + \sqrt{3}}{6}\right),$$

$$w_4(x, y) = \left(\frac{x + 2}{3}, \frac{y}{3}\right). \tag{1.3}$$

The von Koch curve is the attractor of the IFS $\{[0, 1] \times [0, 1]; w_i : i = 1, 2, 3, 4\}$. The "trema" and "dragon" type construction of Koch curves can be found in [4]. The following is the MATLAB code for generating the von Koch curve corresponding to the contractions in Eq. (1.3) (Fig. 1.2).

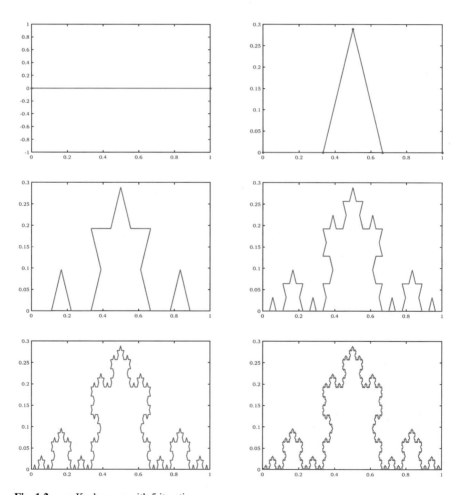

Fig. 1.2 von Koch curve with 5 iterations

```
%Koch  curve
%We  are  taking  transformations  w_i=A_i  x+B_i  for  i=1,2,3,4
%A1=[1/3  0;0  1/3];A2=[1/6  -sqrt(3)/6;sqrt(3)/6  1/6];
%A3=[1/6  sqrt(3)/6;-sqrt(3)/6  1/6];A4=[1/3  0;0  1/3];
%B=[0  1/3  1/2  2/3;0  0  sqrt(3)/6  0];B_i=B(:,i);
clc;clear  all;close  all;
A1=[1/3  0;0  1/3];
A2=[1/6  -sqrt(3)/6;sqrt(3)/6  1/6];
A3=[1/6  sqrt(3)/6;-sqrt*(3)/6  1/6];
A4=[1/3  0;0  1/3];
B=[0  1/3  1/2  2/3;0  0  sqrt(3)/6  0];
X=[0  1;0  0];%  input
x=X(1,:);y=X(2,:);
plot(x,y,'*-');
figure
Termination=5;
for  iter=1:Termination
    if  iter==1
        w1=A1*X+B(:,1);  w2=A2*X+B(:,2);
        w3=A3*X+B(:,3);  w4=A4*X+B(:,4);
        X=[w1  w2  w3  w4];
        X=unique(X','rows');
        x=X(:,1);    y=X(:,2);
        plot(x,y,'*-');
        X=X';
    else  %  more  than  one  iteration
        figure
        w1=A1*X+B(:,1);%First  tranformation
        T1=w1;
        w2=A2*X+B(:,2);%Second  transformation
        T2=w2;
        w3=A3*X+B(:,3);%Third  transformation
        T3=w3;
        w4=A4*X+B(:,4);%Fourth  transformation
        T4=w4;
        X=[T1  T2  T3  T4];
        x=X(1,:);y=X(2,:);
        plot(x,y);
    end
end
```

1.2.3 Dragon Curve

The construction of dragon curve via the line segments is discussed here. For the
first iteration, replace the line segment with two segments, each scaled by a ratio
$r = 1/\sqrt{2}$ such that the original segment would have been the hypotenuse of an
isosceles right triangle. Following along the original segment, two new segments are
placed to the left. For the second iteration, each of the segments are replaced with two
new segments at right angles, each scaled by the ratio r. The new segments are placed
to the left then to the right along the segments of the first iteration. Continuing the

similar construction, always alternating new segments between left and right along the segments of the previous iteration generates the "dragon curve". The fundamental theorem for generating the dragon curve can be seen in [6]. Let X be the line segment joining two points $(0,0),(1,0)$ and consider the IFS on X consistingn following two contractions,

$$w_1(x, y) = \left(\frac{x}{2} - \frac{y}{2}, \frac{x}{2} + \frac{y}{2}\right), \bullet$$
$$w_2(x, y) = \left(-\frac{x}{2} - \frac{y}{2} + 1, \frac{x}{2} - \frac{y}{2}\right). \tag{1.4}$$

The IFS $\{X; w_i : i = 1, 2\}$ generates the dragon curve, where w_1 and w_2 are provided in Eq. (1.4), which is displayed in Fig. 1.3. The associated MATLAB code for generating Fig. 1.3 is provided below.

```
%Dragon curve
%We are taking transformations w_i=A_i x+B_i for i=1,2
%A1=[1/2  -1/2;1/2  1/2];A2=[-1/2  -1/2;1/2  -1/2];
%B=[0 1;1 0];B_i=B(:,i);
clc;clear all;close all;
A1=[1/2  -1/2;1/2  1/2];A2=[-1/2  -1/2;1/2  -1/2];
B=[0 1;1 0];
X=[0 1;0 0];% input
x=X(1,:);y=X(2,:);
plot(x,y,'*-');
figure
Termination=20;
for iter=1:Termination
    if iter==1
        w1=A1*X+B(:,1); w2=A2*X+B(:,2);
        X=[w1 w2];
        X=unique(X','rows');
        x=X(:,1);    y=X(:,2);
        plot(x,y,'*-');
        X=X';
    else % more than one iteration
        figure
        w1=A1*X+B(:,1);%First tranformation
        T1=w1;
        w2=A2*X+B(:,2);%Second transformation
        T2=w2;
        X=[T1 T2];
        x=X(1,:);y=X(2,:);
        plot(x,y);
    end
end
```

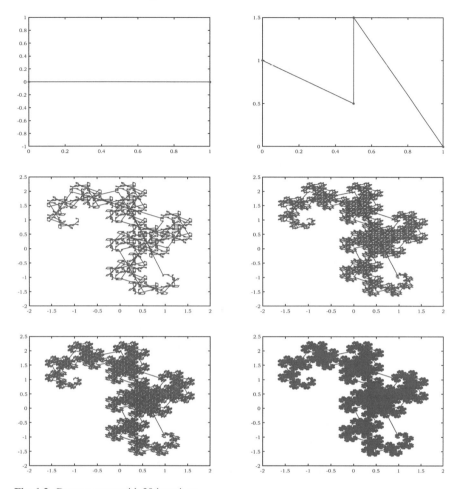

Fig. 1.3 Dragon curve with 20 iterations

1.2.4 Fern Leaf

A fern leaf is a perfect example for both random fractal and deterministic fractal. Though it comes under the classification of random fractals, it can be constructed as a deterministic fractal to a specific IFS containing the following two contraction mappings.

$$w_1(x, y) = \left(0, \frac{4y}{25}\right),$$

$$w_2(x, y) = \left(\frac{17x}{20} + \frac{y}{25}, -\frac{x}{25} + \frac{17y}{20} + \frac{4}{25}\right),$$

$$w_3(x, y) = \left(\frac{x}{5} - \frac{13y}{50}, -\frac{23x}{100} + \frac{11y}{50} + \frac{4}{25}\right),$$

$$w_4(x, y) = \left(-\frac{3x}{20} + \frac{7y}{25}, \frac{13x}{50} + \frac{6y}{25} + \frac{11}{25}\right).$$

(1.5)

Barnsley has discussed the algorithms for constructing fern leaf in [2]. The MATLAB code is provided below to generate a fern leaf utilising contraction maps in Eq. (1.5) (Fig. 1.4).

```
function fern
% Barnsley Fern Fractal generator
%(https://stackoverflow.com/questions/39628601/generating-
    barnsley-fern
-fractal-in-matlab)
n = 10^5;
x = zeros(n,1);
y = zeros(n,1);
for i = 2:n
    r = rand;
    if (0 <= r) && (r < 0.01) %First transformation
        x(i) = 0;
        y(i) = 0.16*y(i-1);
    elseif (0.01 <= r) && (r < 0.86)% Second transformation
        x(i) = 0.85 * x(i-1) + 0.04 * y(i-1);
        y(i) = -0.04 * x(i-1) + 0.85 * y(i-1) + 1.6;
    elseif (0.86 <= r) && (r < 0.93) % Third transformation
        x(i) = 0.2  * x(i-1) - 0.26 * y(i-1);
        y(i) = 0.23 * x(i-1) + 0.22 * y(i-1) + 1.6;
    else  %Fourth transformation
        x(i) = -0.15 * x(i-1) + 0.28 * y(i-1);
        y(i) =  0.26 * x(i-1) + 0.24 * y(i-1) + 0.44;
    end
end
plot(x,y,'.', 'Color', [85, 125, 65]/256, 'markersize', 0.1)
end
```

1.2.5 Sierpinski Carpet

The construction commences with a filled solid square denoted by C_0. Divide C_0 into 9 smaller congruent squares and remove the center square leaving its boundary to get C_1. Next each of eight remaining solid squares are subdivided into 9 congruent squares and the center squares are removed from each to get C_2. Continuing the process, a decreasing sequence of sets $C_0 \supset C_1 \supset C_2 \cdots$ is obtained. The intersection

Fig. 1.4 Fern leaf

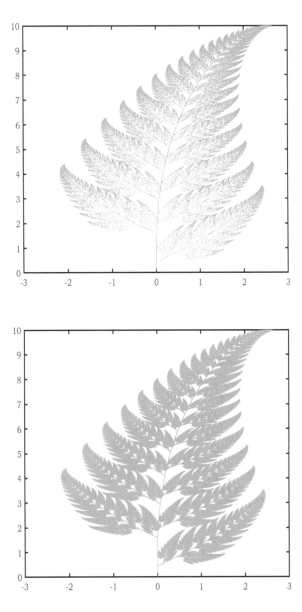

of all the sets in this sequence is the Sierpinski carpet. Let X be a unit square on $\mathcal{R}^2$ with the vertices $A = (0, 0)$, $B = (1, 0)$, $C = (0, 1)$ and $D = (1, 1)$ and the IFS on X consists of the following eight contractions,

$$w_1(x, y) = \left(\frac{x}{3}, \frac{y}{3}\right),$$

$$w_2(x, y) = \left(\frac{x}{3}, \frac{y}{3} + \frac{1}{3}\right),$$

$$w_3(x, y) = \left(\frac{x}{3}, \frac{y}{3} + \frac{2}{3}\right),$$

$$w_4(x, y) = \left(\frac{x}{3} + \frac{1}{3}, \frac{y}{3}\right),$$

$$w_5(x, y) = \left(\frac{x}{3} + \frac{1}{3}, \frac{y}{3} + \frac{2}{3}\right),$$

$$w_6(x, y) = \left(\frac{x}{3} + \frac{2}{3}, \frac{y}{3}\right),$$

$$w_7(x, y) = \left(\frac{x}{3} + \frac{2}{3}, \frac{y}{3} + \frac{1}{3}\right),$$

$$w_8(x, y) = \left(\frac{x}{3} + \frac{2}{3}, \frac{y}{3} + \frac{2}{3}\right).$$

(1.6)

The IFS $\{X; w_i : i = 1, 2, \ldots, 8\}$ with maps w_i in Eq. (1.6) generates Sierpinski carpet as its attractor. The MATLAB code to generate the carpet employing the above contractions is provided as follows (Fig. 1.5).

```
%Sierpinski Carpet
%We are taking transformations w_i=A x+B_i for i=1,2,3,4
%A=[1/3 0;0 1/3];
%B=[0 0 0 1/3 1/3 2/3 2/3 2/3;0 1/3 2/3 0 2/3 0 1/3 2/3]; B_i=B(:,i);
%A=(0,0),B=(1,0),C=(0,1) and D=(1,1);x=(0 1 0 1;0 0 1 1)
clc;clear all;close all;
A=[1/3 0;0 1/3];
B=[0 0 0 1/3 1/3 2/3 2/3 2/3;0 1/3 2/3 0 2/3 0 1/3 2/3];
X=[0 1 1 0;0 0 1 1];% input
X1=[X X(:,1)];
x=X1(1,:);y=X1(2,:);
plot(x,y,'*-');
figure
Termination=5;
for iter=1:Termination
    if iter==1
        w1=A*X+B(:,1);w2=A*X+B(:,2);w3=A*X+B(:,3);w4=A*X+B(:,4);
        w5=A*X+B(:,5);w6=A*X+B(:,6);w7=A*X+B(:,7);w8=A*X+B(:,8);
        X=[w1 w2 w3 w4 w5 w6 w7 w8];
        [m n]=size(X);
        for i=1:n/4
            ivalue=i;
            D=[X(:, 4*(i-1)+1:4*i) X(:,4*(i-1)+1)];
            x=D(1,:);y=D(2,:);
            plot(x,y)
            fill(x,y,'b')
            hold on
        end
    else % more than one iteration
```

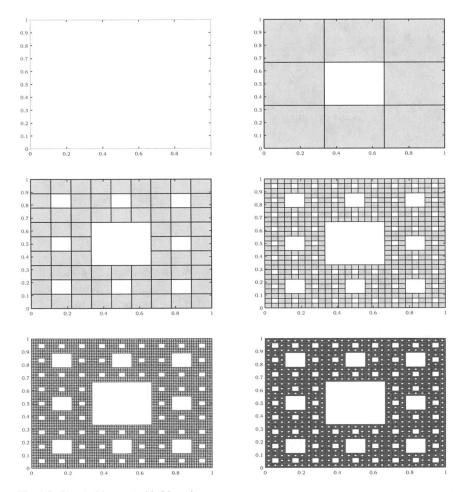

Fig. 1.5 Sierpinski carpet with 5 iterations

```
figure
w1=A*X+B(:,1);%First tranformation
T1=w1;
w2=A*X+B(:,2);%Second transformation
T2=w2;
w3=A*X+B(:,3);%Third transformation
T3=w3;
w4=A*X+B(:,4);%Fourth transformation
T4=w4;
w5=A*X+B(:,5);%Fiveth tranformation
T5=w5;
w6=A*X+B(:,6);%Sixth transformation
T6=w6;
w7=A*X+B(:,7);%Seventh transformation
T7=w7;
```

```
w8=A*X+B(:,8);%Eighth transformation
T8=w8;
X=[T1 T2 T3 T4 T5 T6 T7 T8];
[m n]=size(X);
for i=1:n/4
    D=[X(:, 4*(i-1)+1:4*i) X(:,4*(i-1)+1)];
    x=D(1,:);y=D(2,:);
    fill(x,y,'b')
    hold on
end
    end
end
```

1.3 Fractal Dimensions

Hausdorff dimension can be defined for any set and it is mathematically convenient to manipulate, since it is based on measures. It involves taking the infimum over covers of a given set, say K, consisting of balls of radius less than or equal to $\epsilon > 0$ and this makes its explicit computation difficult. A slight simplification is obtained by considering only covers containing balls of radius equal to $\epsilon > 0$. This gives rise to the concept of box dimension. For interesting results on fractal dimension of fractals and fractal functions, refer the book [7, 8]. The algorithm for box counting method is provided in the following subsection.

1.3.1 Box Counting Algorithm

Box counting is a method of gathering data for analyzing complex patterns by breaking a data set, object, image, etc. into smaller and smaller pieces, typically "box"-shaped, and analyzing the pieces at each smaller scale.

Using the box counting method, fractal dimension is the slope of the line when we plot the value of $\log(N)$ on Y-axis against the value of $\log(r)$ on X-axis. The same equation is used to define the fractal dimension, D. The MATLAB code for box counting algorithm is given as follows.

```
clc;clear all;close all;
c=imread('koch_iter5.jpg');
image(c)
axis image
[n,r] = boxcount(c,'slope')
df = -diff(log(n))./diff(log(r));
disp(['Fractal dimension, Df = ' num2str(mean(df(4:8))) ' +/- '
    num2str
    (std(df(4:8)))]);
```

Output: N = Columns 1 through 8
  16664999 4167500 1042500 260625 65417 16485 4187 1080
  Columns 9 through 14
  280 70 20 6 2 1
  r = Columns 1 through 8
  1 2 4 8 16 32 64 128
  Columns 9 through 14
  256 512 1024 2048 4096 8192
  Fractal dimension, Df = 1.26 +/- 0.02052

1.3.2 Higuchi Algorithm

Higuchi developed a technique to measure the fractal dimension of the data $(x, f(x))$ forming the graph of a function f on the unit interval. The Higuchi method takes a signal, discretized into the form of a time series, $x_0, x_1, x_2, \ldots, x_N$. From given time series, we construct a new time series, X_k^m, defined as:

$$X_k^m : f(m), f(m+k), f(m+2k), \ldots, f\left(m + \lfloor \frac{N-m}{k} \rfloor .k\right),$$

where $\lfloor \frac{N-m}{k} \rfloor$ is the integer part of $\frac{N-m}{k}$, $k \in [1, 2, \ldots, k_{max}]$ indicate the interval time and $m \in [1, 2, \ldots, k]$ is the initial time. We define the length of the curve , X_k^m, as follows,

$$L_m(K) = \sum_{i=1}^{\lfloor \frac{N-m}{k} \rfloor} \frac{N-1}{\lfloor \frac{N-m}{k} \rfloor .k^2} \left\{ \left(|f(m+ik) - f(m+(i-1).k)| \right) \right\}.$$

The length of the curve for the time interval k is then defined as the sum over the k sets of $L_m(k)$,

$$L(k) = \sum_{m=1}^{k} L_m(k).$$

Fractal dimension is the slop of the data $\{(\log \frac{1}{k}, \log L(k))\}$. The MATLAB code for estimating Higuchi fractal dimension is provided below.

```
%Higuchi Fractal Dimension
clc;clear all;close all;
format 'short'
x=[1 5 7 8 9 15 16 12 18 20];
N=length(x);
k_max=5;%Interval time
M=0;
for k=1:k_max
    for m=1:k
        H(k,m)=km(x,k,m,N);
    end
    D(k)=sum(H(k,:));
end
LD=log10(D);
k=1:k_max;
Lk=log10(k);
scatter(Lk,LD,'b','*');
P=polyfit(Lk,LD,1)
slope = P(1)
intercept = P(2);
yfit = P(1)*Lk+P(2);
% P(1) is the slope and P(2) is the intercept
hold on;
plot(Lk,yfit,'r-.')
%%%%
function [H]=km(x,k,m,N)
AA=floor((N-m)/k);
for j=1:AA+1
    aa(j)=m+(j-1)*k;
    HI(j)=x(aa(j));
end
aa;
HI;
for i=1:length(HI)-1
    L(i)=abs(HI(i+1)-HI(i));
end
u=(N-1)/(AA*k^2);
H=(1/u)*sum(L);
end
```

Output: Higuchi Fractal Dimension is 1.0578 for $K_{max} = 5$.

1.3.3 Katz Algorithm

The Katz algorithm is used to compute fractal dimension for signals and it is expressed as

$$D = \frac{\log \frac{L}{a}}{\log \frac{d}{a}}, \tag{1.7}$$

here L is the total length of the curve, d is the diameter (planar extent) of the curve and a is the average distance between two successive points. Length of the curve L can be interpreted as accumulated change of speech signal values and is calculated using simple Euclidean distance formula:

$$L = \sum_{i=1}^{n} l_{i,i+1} = \sum_{i=1}^{n} \sqrt{(x_{i+1} - x_i)^2 + (y_{i+1} - y_i)^2}. \tag{1.8}$$

Planar extent d is equal to the maximum distance between the first point and any other point of the curve:

$$d = \max\{l_{1,i}\}, \ 2 \le i \le n + 1. \tag{1.9}$$

With this description, the MATLAB code for Katz fractal dimension is given as follows.

```
% Katz fractal dimension for given signal values
clc; clear all; close all;
%Akima data {(0, 10), (2, 10), (3, 10), (5, 10),
%(6, 10), (8, 10), (9, 10.5), (11, 15), (12, 50), (14, 60), (15,
    85)}
M=[0    10;2  10;3    10;5  10;6  10;8  10;9  10.5;11   15;12  50;14  60;15
    85];
 [m n]= size (M)
 % Length of the two succesive points
 for  i =1:m-1
      LL( i )= sqrt ((M( i +1 ,1)-M( i ,1)) ^2+(M( i +1 ,2)-M( i ,2)) ^2);
 end
L=sum (LL)%Total  length  of  the  curve
 for  i =1:m
      dd ( i )= sqrt ((M(1 ,1)-M( i ,1)) ^2+(M(1 ,2)-M( i ,2)) ^2);
 end
 d=max ( dd )% Diameter    of  the  curve
 a=L/m % a  is  the  average  distanc
 D= log (L/a)/ log (d/a)%Katz  fractal  dimension
 % Output  is  m =   11;n =2
 % L =     84.2748;  d =  76.4853;  a  = 7.6613
 % D =   1.0422
```

Input Data: Akima data is
{(0, 10), (2, 10), (3, 10), (5, 10), (6, 10), (8, 10), (9, 10.5), (11, 15), (12, 50), (14, 60), (15, 85)}
Output is $m = 11$; $n = 2$,
$L = 84.2748$; $d = 76.4853$; $a = 7.6613$. Fractal dimension, $D = 1.0422$.

References

1. B.B. Mandelbrot, *The Fractal Geometry of Nature*, vol. 1 (WH Freeman, New York, 1982)
2. M.F. Barnsley, *Fractals Everywhere* (Academic Press Inc, Boston, MA, 1988)
3. M.F. Barnsley, Fractal functions and interpolation. Constr. Approx. **2**, 303–329 (1986)
4. G.A. Edgar, *Measure, Topology, and Fractal Geometry*, vol. 2 (Springer, 2008)
5. P.R. Massopust, *Fractal Functions, Fractal Surfaces, and Wavelets* (Academic Press, 2014)
6. S. Banerjee, M.K. Hassan, S. Mukherjee, A. Gowrisankar, *Fractal Patterns in Nonlinear Dynamics and Applications* (CRC Press, 2020)
7. K. Falconer, *Fractal Geometry: Mathematical Foundations and Applications* (Wiley, 2004)
8. S. Banerjee, D. Easwaramoorthy, A. Gowrisankar, *Fractal Functions, Dimensions Signal Analysis* (Springer, 2021)

Chapter 2
Univariate Fractal Functions

The method of fractal interpolation is developed with an ultimate aim of approximating the naturally existing complicated functions which share non-smoothness in their patterns. This method provides a satisfactory generalization of classical interpolation techniques since classical approaches are emerged only to approximate the smooth functions however, most of the real world experimental functions are highly irregular. Like classical interpolation technique, a finite data set is taken in the fractal interpolation scheme and a continuous function is determined such that whose graph passes through all the given finite set of data points. In the case of fractal interpolation, the required continuous function (i.e., fractal function) does not posses any explicit form and its graph is generated as an attractor of a special kind of iterated function system. Yet, Barnsley has utilized the Read-Bajraktarević operator in order to provide a functional equation for the fractal interpolation function in [1]. Since from the emergence of fractal interpolation functions in late 1980s, several types of fractal interpolation functions have been discovered including univariate [2–5], bivariate [6–9] and fractal functions on higher dimensional spaces [10–13]. The following is the description of the construction of univariate fractal interpolation function.

2.1 Affine Fractal Interpolation

Let $N \in \mathbb{N}$. Consider the data set $\{(x_i, y_i) \in \mathbb{R} \times \mathbb{R} : i = 1, 2, \ldots, N+1\}$ with $x_1 < x_2 < \cdots < x_{N+1}$ and x_i's are not required to be equidistant. Consider the closed sub-intervals of real-line $I = [x_1, x_{N+1}]$ and $I_i = [x_i, x_{i+1}]$, $\forall i = 1, 2, \ldots, N$. The graph of continuous function f interpolating the provided data set such that $f(x_i) = y_i$, $\forall i = 1, 2, \ldots, N+1$ is determined as an attractor of the following iterated function system,

$$\{X; w_i : i = 1, 2, \ldots, N\}. \tag{2.1}$$

© The Author(s), under exclusive license to Springer Nature Switzerland AG 2023
S. Banerjee et al., *Fractal Patterns with MATLAB*,
SpringerBriefs in Complexity,
https://doi.org/10.1007/978-3-031-48102-4_2

Where $X := I \times \mathbb{R}$ is a complete metric space with respect to the metric equivalent to Euclidean metric and $w_i : X \to I_i \times \mathbb{R}, i = 1, 2, \ldots, N$ are contraction maps defined by

$$w_i(x, y) = (L_i(x), F_i(x, y)), \; x, y \in X. \tag{2.2}$$

The maps involved in the definition of contraction maps w_i are given by $L_i : I \to I_i$ and $F_i : I \times \mathbb{R} \to \mathbb{R}$. Maps L_i are N homeomorphisms and F_i are continuous mappings satisfying

$$|L_i(x) - L_i(x')| \le c_i|x - x'|,$$
$$|F_i(x, y) - F_i(x, y')| \le r_i|y - y'|.$$

for all $x, x' \in I$, $u, y' \in \mathbb{R}$, $c_i, r_i \in (0, 1)$. It is to be noted that F_i is a contraction with respect to second variable. The maps L_i and F_i obey the following join-up conditions

$$\begin{aligned} L_i(x_1) &= x_i, \\ L_i(x_{N+1}) &= x_{i+1}, \\ F_i(x_1, y_1) &= y_i, \\ F_i(x_{N+1}, y_{N+1}) &= y_{i+1}, \end{aligned} \tag{2.3}$$

as prescribed in [1]. Considering the IFS (2.1), a set-valued map $\mathcal{W}$ defined on the non-empty compact subsets of X, say $\mathcal{H}(X)$, is expressed by

$$\mathcal{W}(C) = \bigcup_{i=1}^{N} w_i(C),$$

for any $C \in \mathcal{H}(X)$. Since X is a complete metric space, the Hausdorff metric completes $\mathcal{H}(X)$. Note that the map W is a finite union of contraction maps w_i and defined on the complete metric space $\mathcal{H}(X)$. Then, by the Banach contraction principle, $\mathcal{W}$ possess a unique invariant compact set, say $\mathcal{G}_f$ satisfying $\mathcal{G}_f = \mathcal{W}(\mathcal{G}_f)$ and it is the graph of a required continuous function $f : I \to \mathbb{R}$ such that $\mathcal{G}_f := \{(x, f(x)) : x \in I\}$. The graph of the function f obtained as an attractor of the IFS (2.1) is referred as a Fractal Interpolation Function (FIF), in short *fractal function*.

Now, let us overview the generation of functional equation (also called as fixed point equation) for the above described fractal interpolation function by means of defining the Read-Bajraktarević (RB) operator. In this generation, the fractal function is shown as a fixed point of the RB operator T. Consider the Banach space of continuous functions, $\mathbb{C}$, such that

$$\mathbb{C} = \{h : I \to \mathbb{R} \mid h \text{ is continuous on } I, h(x_1) = y_1, h(x_{N+1}) = y_{N+1}\},$$

with the sup norm $\|h\|_\infty = \max\{|h(x)| : x \in I\}$. A contraction map is induced on this complete metric space $\mathbb{C}$, defined by

$$T(h(x)) = F_i\left(L_i^{-1}(x), h \circ L_i^{-1}(x)\right), \ x \in I_i, \ i = 1, 2, \ldots, N,$$

The contraction map T admits a unique fixed point, which is the above defined fractal interpolation function f such that $f(x) = T(f(x))$. In addition, it obeys the below given functional equation

$$f(x) = F_i\left(L_i^{-1}(x), f \circ L_i^{-1}(x)\right), \ x \in I_i, \ i = 1, 2, \ldots, N. \tag{2.4}$$

The general IFS employed to study various kinds of fractal interpolation functions is given below,

$$\begin{aligned} L_i(x) &= a_i x + b_i, \\ F_i(x, y) &= \alpha_i y + q_i(x), \ i = 1, 2, \ldots, N. \end{aligned} \tag{2.5}$$

Where $\{\alpha_i : i = 1, 2, \ldots, N\}$ are the free parameters called as vertical scaling factors (also referred as contraction factors) of the contraction maps w_i satisfying $-1 < \alpha_i < 1$ and $q_i : I \to \mathbb{R}$ are continuous functions obeying the conditions

$$\begin{aligned} q_i(x_1) &= y_i - \alpha_i y_1, \\ q_i(x_{N+1}) &= y_{i+1} - \alpha_i y_{N+1}, \ \forall i = 1, 2, \ldots, N. \end{aligned}$$

Among the univariate fractal interpolation function, types of fractal functions can be generated including linear fractal function, quadratic fractal function, alpha fractal function and so on. The continuous function q_i has a prominent role in differentiating and constructing new kinds of fractal functions. For instance, if q_i is taken as a linear function of the form $c_i x + d_i$, then the corresponding IFS generated is as follows

$$\begin{aligned} L_i(x) &= a_i x + b_i, \\ F_i(x, y) &= \alpha_i y + c_i(x) + d_i, \ i = 1, 2, \ldots, N, \end{aligned} \tag{2.6}$$

thus the IFS (2.6) invokes a linear fractal interpolation function. In the similar manner, for each unique q_i, a special kind of fractal interpolation function is generated. Moreover fractal interpolation functions can be classified as affine and non-affine functions by correspondingly choosing the the continuous functions q_i as affine and non-affine. Suppose q_i is of the form,

$$q_i = c_i x + d_i,$$

one can get the affine transformations as follows,

$$w_i \begin{pmatrix} x \\ y \end{pmatrix} = \begin{pmatrix} a_i & 0 \\ c_i & \alpha_i \end{pmatrix} \begin{pmatrix} x \\ y \end{pmatrix} + \begin{pmatrix} b_i \\ d_i \end{pmatrix},$$

where a_i, b_i, c_i, d_i are the real parameters. By predefining the scaling factors α_i and employing the join-up conditions, the following can be determined, for $i = 1, 2. \ldots, N$,

$$
\begin{aligned}
a_i &= \frac{x_{i+1} - x_i}{x_{N+1} - x_1}, \\
b_i &= \frac{x_{N+1} x_i - x_1 x_{i+1}}{x_{N+1} - x_1}, \\
c_i &= \frac{(y_{i+1} - y_i) - \alpha_i (y_{N+1} - y_1)}{x_{N+1} - x_1}, \\
d_i &= \frac{(x_{N+1} y_i - x_1 y_{i+1}) - \alpha_i (x_{N+1} y_1 - x_1 y_{N+1})}{x_{N+1} - x_1}.
\end{aligned}
\tag{2.7}
$$

The fractal function associated with the above defined affine transformation is called as the *affine fractal interpolation function*.

Following is the MATLAB code for generating affine fractal graphs with constant scaling parameters.

```
%% Affine Fractal Interpolation Function
%% L_i (x)=a_i x+b_i
%% F_i(x,y)=alpha_i (x) *y +Q_i(x),
% where Q_i(x) is the affine function
clc;clear all;close all;
format 'short'
x=[0   1/3 1/2 2/3  1];y=[1   3 5/2 3.5 3/2];% Data
iter=input('Enter the number of iterations:=');
lx=length(x);
%alpha =[0 0 0]
alpha =[0.3 0.2 0.1  0.6];
N=lx;
for i=1:lx-1
    diff_x(i)=x(i+1)-x(i);length_x=x(N)-x(1);
    a(i)=diff_x(i)/length_x;
    b(i)=(x(N)*x(i)-x(1)*x(i+1))/length_x;
    c(i)=[y(i+1)-y(i)-alpha(i)*(y(N)-y(1))]/length_x;
    d(i)=[x(N)*y(i)-x(1)*y(i+1)-alpha(i)*(x(N)*y(1)-x(1)*y(N))]/
        length_x;
end
abcd_values=[a' b' c' d']
%%%%%%%%%%%%%%%%%%%%%%%%%%%%%%%%%%
L=[];L1=[];X1=[];Y1=[];X=[];Y=[];
p=N;
for k=1:iter
    for i=1:N-1
        for t1=1:p
            if (k==1)% First iteration
                % Input data is (x,y) or given data
                L(i,t1)=a(i)*x(t1)+b(i);
```

```
                    L1(i,t1)=alpha(i)*y(t1)+c(i)*x(t1)+d(i);
               else % More than one iteration
                    % Input data is (X1,Y1) or output after the first
                      iterration
                    L(i,t1)=a(i)*X1(t1)+b(i);
                    L1(i,t1)=alpha(i)*Y1(t1)+c(i)*X1(t1)+d(i);
               end
          end
          X=[X L(i,:)];
          Y=[Y L1(i,:)];
     end
     X1=X;     Y1=Y;     X=[];     Y=[];     g=[X1' Y1'];
     g=str2num(num2str(g,10));g=unique(g,'rows');
     X1=g(:,1);Y1=g(:,2);     [X1 Y1]; p=length(X1);
     %%%
end
plot(x,y,'.k','markersize',20);hold on
plot(X1,Y1,'r-');title('AffineFIF_First Iteration ')
% %%%%%%%%%%%%%%%%%%%%%%%%%%%%%%%%%%%
%
```

2.1.1 *Vertical Scaling Factors*

The scale vector $\alpha = (\alpha_1, \alpha_2, \ldots, \alpha_N)$ associated with the IFS (2.6) has a significant dominance in determining fractal behaviours of the fractal function, which includes fractal dimension, shape preserving as well as shape modifying properties. Further, the scaling factors govern the closeness of fit corresponding to the provided data set. In general, vertical scaling factors are utilized to modify the shape and size of the curves along the vertical direction. While dealing fractal interpolation functions, the absolute value of vertical scaling factors is constrained to be less than 1 and thus they contracts the fractal curves according to given values of the scale vector α. Hence, they are also called as contractivity factors for the contractions w_i.

For the choice of vertical scaling factors as constants, the naturally existing self-similar functions are better approximated, since if the curves are strictly self-similar, the constant vertical parameters easily makes the same ratio of compression in each sub-interval to yield closer fit. On the contrary, if the curves show less self-similarity, the constant scalings may lead to loss of flexibility and cause more approximation errors. To address this issue, fractal interpolation function with function (variable) scaling factors have been introduced to fit the non-stationary data set. In [14], the vertical scaling parameters α_i are chosen as continuous functions on the closed interval I (i.e.,)

$$\alpha_i : I \to [0, 1), \ \forall\, i = 1, 2, \ldots, N,$$

satisfying

$$\|\alpha\|_\infty = \sup\{\|\alpha_i\|_\infty : i = 1, 2, \ldots, N\} < 1$$

and the following IFS is obtained

$$L_i(x) = a_i x + b_i,$$
$$F_i(x, y) = \alpha_i(x)y + q_i(x), \ i = 1, 2, \ldots, N. \tag{2.8}$$

The fractal function generated using the IFS (2.8) is referred as the fractal interpolation function with variable scaling factors.

2.1.2 Affine Fractal Function with Variable Scaling

Consider the IFS (2.8) with the following continuous function,

$$q_i(x) = c_i x + d_i, \ \forall \, i = 1, 2, \ldots, N. \tag{2.9}$$

Thus, the IFS for generating the affine fractal interpolation function with variable scaling is obtained as

$$L_i(x) = a_i x + b_i, \ F_i(x, y) = \alpha_i(x)y + c_i x + d_i, \ \forall \, i = 1, 2, \ldots, N.$$

The MATLAB code for generating affine fractal interpolation function with variable scaling is provided below.

```
%Affine Fractal Interpolation Function   with variable scaling
% L_i (x)=a_i x+b_i
% F_i(x,y)=alpha_i (x) *y +Q_i(x),
% where Q(x) is the affine function
clc;clear all;close all;
format 'short'
x=[0 1/4 1/2 3/4 1];y=[1  3  5/2 3.5 3/2];% Data
iteration=1;
[X1,Y1]=VAR_affine(x,y,iteration);
subplot(2,2,1)
plot (x,y,'.b','markersize',20);%Plotting original data
hold on
plot(X1,Y1,'r-','LineWidth',1);%Plotting original and new data
xlabel('Var@AffineFIF:1st iteration')
iteration=3;
subplot(2,2,2)
[X2,Y2]=VAR_affine(x,y,iteration);
plot (x,y,'.b','markersize',20);
hold on
plot(X2,Y2,'r-','LineWidth',1);
xlabel('Var@AffineFIF:3rd iteration')
iteration=8;
subplot(2,2,3)
[X3,Y3]=VAR_affine(x,y,iteration);
plot (x,y,'.b','markersize',20);
hold on
```

```
plot(X3,Y3,'r-','LineWidth',1);
xlabel('Var@AffineFIF: 8th iteration')
iter=8;alpha=[0 0 0 0];
subplot(2,2,4)
[X4,Y4]=AffineFIF(x,y,alpha,iteration);
plot (x,y,'.b','markersize',20);
hold on
plot(X4,Y4,'r-','LineWidth',1);
xlabel('Classical AffineFIF: 8th iteration')
function [X1 Y1]=VAR_affine(x,y,iteration)
N=length(x);lx=N;
a=zeros(lx-1);b=zeros(lx-1);
for i=1:lx-1
    a(i)=(x(i+1)-x(i))/(x(lx)-x(1));
    b(i)=((x(i)*x(lx))-(x(i+1)*x(1)))/(x(lx)-x(1));
end
L=[];L1=[];X1=[];Y1=[];X=[];Y=[];
p=N;
for k=1:iteration
    for i=1:lx-1
        for t1=1:p
            if (k==1)
                L(i,t1)=(a(i)*x(t1))+b(i);
                if(i==1)
                    alpha(i,t1)=x(t1)/(x(lx)-x(1));
                elseif(i==2)
                    alpha(i,t1)=abs(log(1/(x(t1)+2)))/x(lx);
                elseif(i==3)
                    alpha(i,t1)=cos(1-x(t1));
                elseif(i==4)
                    alpha(i,t1)=(sin(x(t1))/(2*(x(lx)-x(1))));
                end
                % co-efficients
                c(i)=[y(i+1)-y(i)-alpha(i,t1)*(y(N)-y(1))]/(x(lx)
                    -x(1));
                d(i)=[x(N)*y(i)-x(1)*y(i+1)-alpha(i,t1)*(x(N)*y
                    (1)-x(1)*y(N))]/(x(lx)-x(1));
                L1(i,t1)=(alpha(i,t1)*y(t1))+c(i)*x(t1)+d(i);
            else
                L(i,t1)=(a(i)*X1(t1))+b(i);
                if(i==1)
                    alpha(i,t1)=X1(t1)/(X1(p)-X1(1));
                elseif(i==2)
                    alpha(i,t1)=abs(log(1/(X1(t1)+2)))/X1(p);
                elseif(i==3)
                    alpha(i,t1)=cos(1-(X1(t1)));
                elseif(i==4)
                    alpha(i,t1)=sin(X1(t1)-1)/(2*(x(lx)-x(1)));
                end
                c(i)=[y(i+1)-y(i)-alpha(i,t1)*(y(N)-y(1))]/(x(lx)
                    -x(1));
                d(i)=[x(N)*y(i)-x(1)*y(i+1)-alpha(i,t1)*(x(N)*y
                    (1)-x(1)*y(N))]/(x(lx)-x(1));
                L1(i,t1)=(alpha(i,t1)*Y1(t1))+c(i)*X1(t1)+d(i);
            end
```

```
            end
        X=[X  L(i ,:) ];
        Y=[Y  L1(i ,:) ];
    end
    X1=X;        Y1=Y;
    X =[];        Y =[];
    p=length(X1);
end
end
%Affine  FIF  with  constant  scaling
function  [X1,Y1]= AffineFIF (x ,y, alpha , iteration )
lx =length(x);
N=lx ;
for  i =1: lx −1
    diff_x ( i )=x ( i +1)−x ( i ); length_x =x (N)−x (1) ;
    a( i )= diff_x ( i )/ length_x ;
    b( i )=(x (N)∗x ( i )−x (1)∗x ( i +1))/ length_x ;
    c ( i )=[y ( i +1)−y ( i )−alpha ( i )∗(y (N)−y (1) ) ]/ length_x ;
    d( i )=[x (N)∗y ( i )−x (1)∗y ( i +1)−alpha ( i )∗(x (N)∗y (1)−x (1)∗y (N)) ]/
        length_x ;
end
abcd_values =[a ' b ' c ' d '];
L =[]; L1 =[]; X1 =[]; Y1 =[]; X =[]; Y =[];
p=N;
for  k =1: iteration
    for  i =1:N−1
        for  t1 =1:p
            if  (k ==1)% First  iteration
                % Input  data  is  (x ,y)  or  given  data
                L( i , t1 )=a ( i )∗x ( t1 )+b( i );
                L1( i , t1 )=alpha ( i )∗y ( t1 )+c ( i )∗x ( t1 )+d( i );
            else % More  than  one  iteration
                % Input  data  is  (X1 ,Y1)  and  output  after  the
                    first  iteration
                L( i , t1 )=a ( i )∗X1( t1 )+b( i );
                L1( i , t1 )=alpha ( i )∗Y1( t1 )+c ( i )∗X1( t1 )+d( i );
            end
        end
        %Concatenation
        X=[X  L(i ,:) ];
        Y=[Y  L1(i ,:) ];
    end
    X1=X;       Y1=Y;       X =[];       Y =[];       g =[X1 ' Y1 '];
    %Removing  duplicates
    g=str2num ( num2str (g ,10) ); g=unique (g , 'rows ');
    X1=g (: ,1) ; Y1=g (: ,2) ;       [X1  Y1 ];  p=length(X1);
end
end
```

2.1.3 Numerical Simulation

Let $\{(0,\ 1), (0.33,\ 3), (0.5,\ 2.5)(0.6,\ 3.5), (1,\ 0.752)\}$ be the given set of interpolation data with vertical scaling factors $\alpha = (0.3, 0.2, 0.1, 0.6)$. The affine FIF corresponding to the given data set for first iteration is graphically shown in Fig. 2.1a. Figure 2.1b is generated by increasing the iteration to three. Figure 2.1c illustrates the affine FIF for eighth iteration. The classical affine FIF is graphically represented in Fig. 2.1d for the choice of zero scaling vector. Consider the variable scaling factor

$$\alpha(x) = \left(\frac{x}{x_N - x_1},\ \left| \frac{x+2}{20x_N} \right|,\ \cos(1-x),\ \frac{\sin(x)}{2(x_N - x_1)} \right)$$

for the same data set. The graph of affine fractal interpolation function with the provided variable scaling factor is generated using the MATLAB code in Sect. 2.1.2. Its graphical illustration is given in Fig. 2.2a. Figure 2.2b and c respectively represents the graph of affine FIF with variable scalings for the third and eighth iteration. With the choice $\alpha_i = [0]_{1\times4}$, classical affine fractal interpolation function is generated and its graph is provided in Fig. 2.2d.

2.2 α-Fractal Interpolation

The α-fractal function is an example of univariate non-affine fractal interpolation function. Unlike the linear fractal interpolation function, a given continuous function g is approximated using α-fractal function g^α. The given continuous function g is generally referred as seed function or germ function and g^α yields a family of fractal functions to each given g and it is referred as the fractal perturbation of g in [15].

Navascués has discovered the α-fractal function which is introduced by Barnsley [1] to provide a fractal analogue for any continuous function. Suppose $g \in \mathcal{C}(I)$, consider the below defined special type of continuous function

$$q_i(x) = g \circ L_i(x) - \alpha_i b(x), \tag{2.10}$$

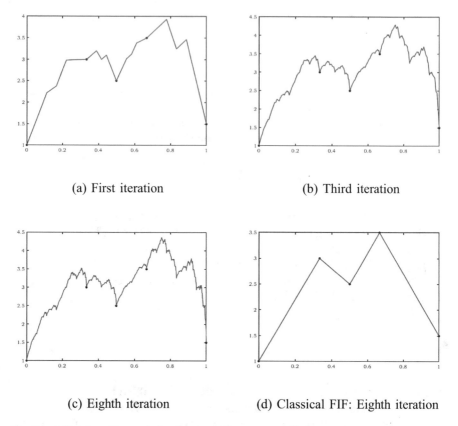

(a) First iteration   (b) Third iteration

(c) Eighth iteration   (d) Classical FIF: Eighth iteration

Fig. 2.1 Affine fractal interpolation function with constant scaling

where b is a base function usually it is a real-valued continuous map equivalent to the germ function g at the end points and $b \neq g$. The IFS corresponding to q_i in Eq. (2.10) is given below

$$L_i(x) = a_i x + b_i,$$
$$F_i(x, y) = \alpha_i y + g \circ L_i(x) - \alpha_i b(x), \quad i = 1, 2, \dots, N, \qquad (2.11)$$

The fractal function associated with the IFS (2.11) is called as the α-fractal interpolation function (in short, α-fractal function) with respect to base function b and the partition $D = x_1 < x_2 < \cdots < x_{N+1}$ of I. Then the α-fractal function satisfies the following functional equation,

$$g^\alpha(x) = g(x) + \alpha_i[(g^\alpha - b) \circ L_i^{-1}(x)], \quad \forall x \in I_i, \ i = 1, 2, \dots, N. \qquad (2.12)$$

For the proper choice of scaling factors, the α-fractal function g^α coincides with the germ function g. If each of the vertical scaling factors α_i are taken as zero,

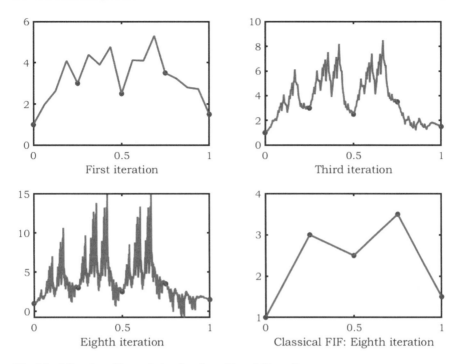

Fig. 2.2 Affine fractal interpolation function with variable scaling

then $g^\alpha = g$. If the scaling factors are taken as continuous functions $\alpha_i(x)$ then the functional equation of the α-fractal function becomes

$$g^\alpha(x) = g(x) + \alpha_i(L_i^{-1}(x))[(g^\alpha - b) \circ L_i^{-1}(x)], \ \forall x \in I_i, \ i = 1, 2, \ldots, N.$$

$$(2.13)$$

The g^α present in the Eq. (2.13) is called the α-fractal function with function scaling factors. The Matlab code for generating α-fractal interpolation function with constant scalings is given below.

```
% $\alpha$-fractal interpolation function
% Base function b(x)=x^3;
clc;clear all;close all;
format 'short';
Fx=inline('x.^3');%Base function
x1=[0  1/4  1/2  3/4  1];y1=Fx(x1);
Data=[x1'  y1'];%Data
iteration=1;
alpha=[0.2  0.3  0.2  0.1];
subplot(2,2,1);
[X1 Y1]=Alphafractal1(iteration,alpha);
plot(x1,y1,'.b','markersize',20);
```

```
hold on
plot(X1,Y1,'r-','LineWidth',1);
xlabel('\alpha-FIF: First iteration')
iteration=3;
[X2 Y2]=Alphafractal1(iteration,alpha);
subplot(2,2,2);
plot (x1,y1,'.b','markersize',20);
hold on
plot(X2,Y2,'r-','LineWidth',1);
xlabel('\alpha-FIF: Third iteration')
subplot(2,2,3);
iteration=8;
[X3 Y3]=Alphafractal1(iteration,alpha);
plot (x1,y1,'.b','markersize',20);%Plotting original data
hold on
plot(X3,Y3,'r-','LineWidth',1);%Plotting original and new data
xlabel('\alpha-FIF: Eighth iteration')
%Classical case
subplot(2,2,4);
iteration=8;alpha=[0 0 0 0] ;
[X4 Y4]=Alphafractal1(iteration,alpha);
plot (x1,y1,'.b','markersize',20);
hold on
plot(X4,Y4,'r-','LineWidth',1);
xlabel('Classical \alpha-FIF: Eighth iteration')
function [X1 Y1]=Alphafractal1(iteration,alpha)
Fx=inline('x.^3');%Base function
x1=[0 1/4 1/2 3/4 1];y1=Fx(x1);
Data=[x1' y1'];%Data
lx=length(x1);
base=inline('x');
a=zeros(1,lx-1);b=zeros(1,lx-1);
for i=1:lx-1
    delta(i)=(y1(i+1)-y1(i))/(y1(lx)-y1(1));
    a(i)=(x1(i+1)-x1(i))/(x1(lx)-x1(1));
    b(i)=((x1(i)*x1(lx))-(x1(i+1)*x1(1)))/(x1(lx)-x1(1));
end
L=[];L1=[];X1=[];Y1=[];X=[];Y=[];
p=lx;
for k=1: iteration
    for i=1:lx-1
        for t=1:p
            if (k==1)%First iteration
                L(i,t)=(a(i)*x1(t))+b(i);
                L1(i,t)=Fx(L(i,t))+alpha(i)*(y1(t)-base(x1(t)));
            else % Morethan one iteration
                L(i,t)=(a(i)*X1(t))+b(i);
                L1(i,t)=Fx(L(i,t))+(alpha(i))*(Y1(t)-base(X1(t)))
                ;
            end
        end
        X=[X L(i,:)];
        Y=[Y L1(i,:)];
    end
    X1=X;    Y1=Y;
```

```
X=[];        Y=[];
%Removing duplicates
g=[X1' Y1'];g=str2num(num2str(g,10));g=unique(g,'rows');
X1-g(:,1);Y1=g(:,2);
p=length(X1);
end
end
```

2.2.1 α-Fractal Function with Variable Scaling

The IFS for generating α-fractal interpolation function with variable vertical scaling factors is provided as

$$L_i(x) = a_i x + b_i,$$
$$F_i(x, y) = \alpha_i(x)y + g \circ L_i(x) - \alpha_i(x)b(x), \ i = 1, 2, \ldots, N,$$

here $\alpha_i : I \to (0, 1)$. The MATLAB code for generating α-fractal function with variable scalings is provided below.

```
%$\alpha$-fractal function with variable scaling
% Base function b(x)=x^3;
clc;clear all;close all;
format 'short';
x=[0 1/4 1/2 3/4 1];
Fx=inline('x.^3');%Base function
y=Fx(x);
iteration=1;
subplot(2,2,1);
[X1 Y1]=Var_alphaFIF(x,iteration);
plot (x,y,'.b','markersize',20);
hold on
plot(X1,Y1,'r-','LineWidth',1);
xlabel('\alpha-FIF: First iteration')
iteration=2;
[X2 Y2]=Var_alphaFIF(x,iteration);
subplot(2,2,2);
plot (x,y,'.b','markersize',20);
hold on
plot(X2,Y2,'r-','LineWidth',1);
xlabel('\alpha-FIF: Second iteration')
subplot(2,2,3);
iteration=3;
[X3 Y3]=Var_alphaFIF(x,iteration);
plot (x,y,'.b','markersize',20);%Plotting original data
hold on
plot(X3,Y3,'r-','LineWidth',1);%Plotting original and new data
xlabel('\alpha-FIF: Third iteration')
subplot(2,2,4);
```

```
iteration =4;
[X4 Y4]= Var_alphaFIF(x,iteration);
plot (x,y,'.b','markersize',20);
hold on
plot(X4,Y4,'r-','LineWidth',1);
xlabel('\alpha-FIF: Fourth iteration')
function [X1 Y1]= Var_alphaFIF(x,iteration)
Fx=inline('x.^3');%Base function
y=Fx(x);
base=inline('x');
lx=length(x);
a=zeros(1,lx-1);
b=zeros(1,lx-1);
A=zeros(1,lx-1);
for i=1:lx-1
    delta(i)=(y(i+1)-y(i))/(y(lx)-y(1));
    a(i)=(x(i+1)-x(i))/(x(lx)-x(1));
    b(i)=((x(i)*x(lx))-(x(i+1)*x(1)))/(x(lx)-x(1));
end
L=[];L1=[];X1=[];Y1=[];X=[];Y=[];
p=lx ;
for k=1: iteration
    for i =1:lx-1
        for t=1:p
            if (k==1)
                L(i,t)=(a(i)*x(t))+b(i);
                if(i==1)
                    alpha(i,t)=0.2+x(t)/(3*(x(lx)-x(1)));
                elseif(i==2)
                    alpha(i,t)=(1/100)*abs(log(1/(x(t)+4)));
                elseif(i==3)
                    alpha(i,t)=(1/2)*cos(1-2*x(t));
                elseif(i==4)
                    alpha(i,t)=0.4+(sin(x(t))/(3*(x(lx)-x(1))));
                end
                L1(i,t)=Fx(L(i,t))+alpha(i)*(y(t)-base(x(t)));
            else
                L(i,t)=(a(i)*X1(t))+b(i);
                if(i==1)
                    alpha(i,t)=0.2+X1(t)/(3*(X1(lx)-X1(1)));
                elseif(i==2)
                    alpha(i,t)=(1/100)*abs(log(1/(X1(t)+4)))/X1(lx);
                elseif(i==3)
                    alpha(i,t)=(1/2)*cos(1-2*X1(t));
                elseif(i==4)
                    alpha(i,t)=0.4+(sin(X1(t))/(3*(X1(lx)-X1(1))));
                end

                L1(i,t)=Fx(L(i,t))+(alpha(i))*(Y1(t)-base(X1(t)));
            end
        end
        X=[X L(i,:) ];
        Y=[Y L1(i,:) ];
    end
    X1=X;
    Y1=Y;
    X=[];
```

```
        Y = [ ] ;
        p = l e n g t h ( X1 ) ;
end
end
%
```

2.2.2 Numerical Simulation

Let $g(x) = x^3$ and $b(x) = x$. Consider the interpolation data $\{0, 0.25, 0.5, 0.75, 1\}$ with the scaling factor $\alpha = (0.2, 0.3, 0.2, 0.1)$ satisfying the constraint $|\alpha_i| < 1$. An α-fractal interpolation function g^α is constructed corresponding to g and b. The graphs of g^α are generated by changing the iteration (first, third and eighth iteration) and portrayed in Fig. 2.3a, b and c. If $\alpha = (0, 0, 0, 0)$, one can obtain the classical alpha fractal function as demonstrated in Fig. 2.3d.

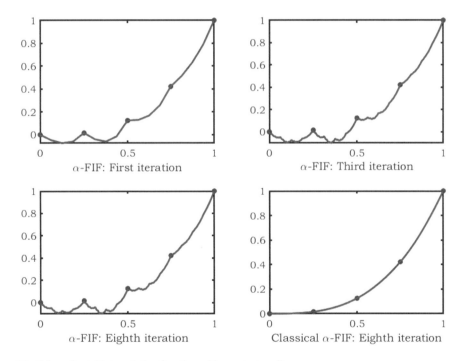

Fig. 2.3 α-fractal interpolation function with constant scaling

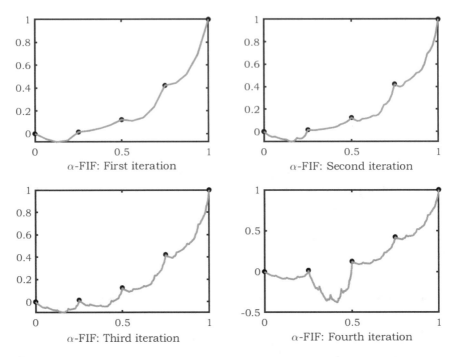

Fig. 2.4 α-fractal interpolation function with variable scaling

If the scaling factors are taken as variables instead of constants,

$$\alpha(x) = \left(\frac{0.2 + x}{3(x_N - x_1)}, \quad \frac{1}{100}\left| \log 1x + 4 \right|, \quad \frac{1}{2}\cos(1 - 2x), \quad 0.4 + \sin\left(\frac{x}{3(x_N - x_1)} \right) \right).$$

Then the corresponding α-fractal interpolation functions are demonstrated in Fig. 2.4a–d.

2.3 Hidden Variable Fractal Interpolation

The given data set $\{(x_i, y_i) : 1 = 1, 2, \ldots, N + 1\}$ is extended from $\mathbb{R}^2$ to $\mathbb{R}^3$ with the inclusion of hidden variables $\{z_i : i = 1, 2, \ldots, N + 1\}$. The new data set $\{(x_i, y_i, z_i) : i = 1, 2, \ldots, N + 1\}$ in $\mathbb{R}^3$ is interpolated using the attractor associated with an IFS $\{\mathbb{R}^3; w_i : i = 1, 2, \ldots, N\}$ consisting of the following maps:

$$w_i \begin{pmatrix} x \\ y \\ z \end{pmatrix} = \begin{pmatrix} a_i & 0 & 0 \\ b_i & \alpha_i & \beta_i \\ c_i & 0 & \gamma_i \end{pmatrix} \begin{pmatrix} x \\ y \\ z \end{pmatrix} + \begin{pmatrix} k_i \\ l_i \\ m_i \end{pmatrix}$$

$$= \begin{pmatrix} L_i(x) \\ F_i(x, y, z) \end{pmatrix},$$

where,

$$L_i(x) = a_i x + b_i,$$
$$F_i(x, y, z) = (\alpha_i y + \beta_i z + p_i(x), \gamma_i z + q_i(x)), \ \forall \ i = 1, 2, \ldots, N + 1,$$

satisfying the endpoint conditions

$$w_i(x_1, y_1, z_1) = (x_i, y_i, z_i) \text{ and } w_i(x_{N+1}, y_{N+1}, z_{N+1}) = (x_{i+1}, y_{i+1}, z_{i+1}). \quad (2.14)$$

The matrix $A_i = \begin{pmatrix} \alpha_i & \beta_i \\ 0 & \gamma_i \end{pmatrix}$ contracts the fractal function on choosing each parameters $\alpha_i, \beta_i, \gamma_i$ satisfying the constraint $\|A_i\| < 1$. The remaining parameters of the map w_i can be determined using the Eq. (2.14), for more information see [16]. Now the attractor of the IFS in $\mathbb{R}^3$ is projected onto $\mathbb{R}^2$ and it is the graph of $f : [x_1, x_{N+1}] \to \mathbb{R}^2$ such that $f(x_i) = (y_i, z_i), \ \forall \ i = 1, 2, \ldots, N + 1$. The hidden variable fractal interpolation function for $\{(x_i, y_i) : i = 1, 2, \ldots, N + 1\}$ is the first component of $f = (f_1, f_2)$ and is defined as the continuous function $f_1 : [x_1, x_{N+1}] \to \mathbb{R}$. The second component f_2 is known as the fractal function interpolating $\{(x_i, z_i) : i = 1, 2, \ldots, N + 1\}$. The MATLAB code for generating hidden variable FIFs is provided as follows.

```
%% Hidden variable fractal interpolation function
%% L_i (x)=a_i x+b_i
%% F_i(x,y,z)=(alpha_iy+beta_iz+p_i(x),gamma_iz+q_i(x)),
% where p_i(x) and q_i(x) are linear polynomials
clc;clear all;close all;
format 'short'
x=[0 1/3 1/2 1];
y=[0 1/3 1/3 1];
z=[0 2/3 1/6 1];% Data
x1=x;
y1=y;
z1=z;
lx=length(x);
n=lx;
alpha=[0.4 0.4 0.4]; beta=[0.3 0.3 0.3]; gamma=[0.4 0.4 0.4];
iteration=10;
 for i=1:lx-1
```

```
a(i)=(x(i+1)-x(i))/(x(n)-x(1));
b(i)=((x(i)*x(n))-(x(i+1)*x(1)))/(x(n)-x(1));
c(i)=((y(i+1)-y(i))-(alpha(i)*(y(n)-y(1)))-(beta(i)*(z(n)-z
     (1))))/(x(n)-x(1));
e(i)=((z(i+1)-z(i))-(gamma(i)*(z(n)-z(1))))/(x(n)-x(1));
d(i)=y(i)-c(i)*x(1)-alpha(i)*y(1)-beta(i)*z(1);
f(i)=z(i)-e(i)*x(1)-gamma(i)*z(1);
c1(i)=((y(i+1)-y(i))-(alpha(i)*(y(n)-y(1))))/(x(n)-x(1));
end
m=n;
for i=1:iteration
    for j=1:lx-1
        for k=1:m
            l(j,k)=(a(j)*x(k)+b(j));
            f1(j,k)=alpha(j)*y(k)+beta(j)*z(k)+c(j)*x(k)+d(j);
            f2(j,k)=gamma(j)*z(k)+e(j)*x(k)+f(j);

        end
    end
    x=reshape(l.',1,numel(l));
    y=reshape(f1.',1,numel(f1));
    z=reshape(f2.',1,numel(f2));
    m=length(x);
end
figure, plot(x,y,'r',x1,y1,'b.','markersize',20)
figure, plot(x,z,'r',x1,z1,'b.','markersize',20)
```

2.3.1 Numerical Simulation

Let a dataset $\{(0, 0, 0), (1/3, 1/3, 2/3), (1/2, 1/3, 1/6), (1, 1, 1)\}$ be given. The scaling parameters are chosen as $\alpha = (0.4, 0.4, 0.4)$, $\beta = (0.3, 0.3, 0.3)$, $\gamma = (0.4, 0.4, 0.4)$ such that $|\alpha_i| < 1$ and $|\beta_i| + |\gamma_i| < 1$ The graphical representations of non-self-affine fractal function f_1 and self-affine FIF f_2 are depicted in Fig. 2.5. The graphs for generated for third iteration and illustrated in Fig. 2.5a and b, whereas Fig. 2.5c and d represent the graphs at tenth iteration.

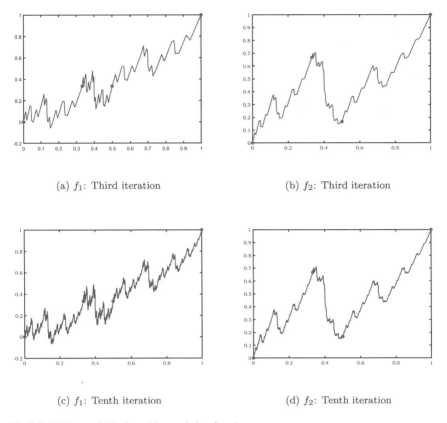

(a) f_1: Third iteration   (b) f_2: Third iteration

(c) f_1: Tenth iteration   (d) f_2: Tenth iteration

Fig. 2.5 Hidden variable fractal interpolation function

References

1. M.F. Barnsley, Fractal functions and interpolation. Constr. Approx. **2**, 303–329 (1986)
2. M.A. Navascués, A fractal approximation to periodicity. Fractals **14**(4), 315–325 (2006)
3. N. Vijender, Bernstein fractal trigonometric approximation. Acta Appl. Math. **159**(1), 11–27 (2019)
4. R. Pasupathi, A.K.B. Chand, M.A. Navascués, Cyclic Meir-Keeler contraction and its fractals. Numer. Funct. Anal. Optim. **42**(9), 1053–1072 (2021)
5. D.-C. Luor, Reproducing kernel Hilbert spaces of fractal interpolation functions for curve fitting problems. Fractals **30**(03), 1–10 (2022)
6. L. Dalla, Bivariate fractal interpolation functions on grids. Fractals **10**(01), 53–58 (2002)
7. A.K.B. Chand, G.P. Kapoor, Hidden variable bivariate fractal interpolation surfaces. Fractals **11**(03), 277–288 (2003)
8. Ri. Song-Il, A new nonlinear bivariate fractal interpolation function. Fractals **26**(04), 1850054 (2018)
9. P. Bouboulis, L. Dalla, V. Drakopoulos, Construction of recurrent bivariate fractal interpolation surfaces and computation of their box-counting dimension. J. Approx. Theory **141**(2), 99–117 (2006)

10. M.F. Barnsley, J. Elton, D. Hardin, P. Massopust, Hidden variable fractal interpolation functions. SIAM J. Math. Anal. **20**(5), 1218–1242 (1989)
11. P.R. Massopust, Vector-valued fractal interpolation functions and their box dimension. Aequationes Mathematicae **42**, 1–22 (1991)
12. H.-Y. Wang, Sensitivity analysis for hidden variable fractal interpolation functions and their moments. Fractals **17**(02), 161–170 (2009)
13. C.-H. Yun, Hidden variable recurrent fractal interpolation functions with function contractivity factors. Fractals **27**(07), 1950113 (2019)
14. H.-Y. Wang, J.-S. Yu, Fractal interpolation functions with variable parameters and their analytical properties. J. Approx. Theory **175**, 1–18 (2013)
15. M.A. Navascués, Fractal polynomial interpolation. Z. Anal. Anwendungen **24**(2), 401–418 (2005)
16. S. Banerjee, D. Easwaramoorthy, A. Gowrisankar, *Fractal Functions, Dimensions and Signal Analysis* (Springer, Berlin, 2021)

Chapter 3
Differentiable Fractal Functions

This chapter presents the fractal patterns for types of fractal splines together with numerical simulations. In [1], Barnsley has discussed the general construction of fractal splines as follows. Consider $L_i(x) = a_i x + b_i$ and $F_i(x, y) = \alpha_i y + q_i(x)$ for $i = 1, 2, \ldots, N$ with $x_1 < x_2 < \cdots < x_{N+1}$ satisfying Eq. (2.3). If $|\alpha_i| < a_i^k$, $k > 0$, $q_i \in \mathcal{C}^k[x_1, x_{N+1}]$ and

$$F_{i,r}(x, y) = \frac{\alpha_i x + q_i^{(r)}(x)}{a_i^r}, \tag{3.1}$$

$$y_{1,r} = \frac{q_1^r(x_1)}{a_1^r - \alpha_1}, \quad y_{N,r} = \frac{q_N^r(x_{N+1})}{a_N^r - \alpha_N}, \quad \text{for } r = 1, 2, \ldots, k. \tag{3.2}$$

Suppose

$$F_{i,r}(x_{N+1}, y_{N+1,r}) = F_{i+1,r}(x_1, y_{1,r}), \quad i = 2, 3, \ldots, N, \ r = 1, 2, \ldots, k,$$

then the IFS $\{(L_i(x), F_i(x, y) : i = 1, 2, \ldots, N\}$ determines $f \in \mathcal{C}^k[x_1, x_{N+1}]$ and $f^{(r)}$ is the FIF associated with $\{(L_i(x), F_{i,r}(x, y) : i = 1, 2, \ldots, N, \ r = 1, 2, \ldots, k\}$.

3.1 Hermite Cubic Fractal Spline

Let $\{(x_i, y_i, d_i) : i = 1, 2 \ldots, N + 1\}$ be the data set with $x_1 < x_2 < \cdots < x_{N+1}$, where y_i denote the function values and d_i represent the derivatives, at the points x_i. In [2], a $\mathcal{C}^1$-cubic Hermite FIF f is constructed with the constraint $|\alpha_i| < sa_i < 1$ for $0 < s < 1$ employing the above-described theory of fractal splines. Let $\mathcal{G} = \{g \in \mathcal{C}^1(I, \mathbb{R}) | g(x_1) = y_1, g(x_{N+1}) = y_{N+1}, g^{(1)}(x_1) = d_1, g^{(1)}(x_{N+1}) = d_{N+1}\}$. The metric induced by the $\mathcal{C}^1$ norm, given by $\|g\|_{\mathcal{C}^1} = \max\{\|g\|_\infty, \|g^{(1)}\|_\infty\}$, is denoted

© The Author(s), under exclusive license to Springer Nature Switzerland AG 2023
S. Banerjee et al., *Fractal Patterns with MATLAB*,
SpringerBriefs in Complexity,
https://doi.org/10.1007/978-3-031-48102-4_3

by ρ. Then the metric space $(\mathcal{G}, \rho)$ is complete. The Read-Bajraktarević (RB) operator $\mathcal{T}$ is defined on $(\mathcal{G}, \rho)$ as

$$(\mathcal{T}g)(x) = \alpha_i g(L_i^{-1}(x)) + q_i(L_i^{-1}(x)),$$

where q_i is a suitable cubic polynomial satisfying $q_i(x_1) = y_i - \alpha_i y_1$ and $q_i(x_{N+1}) = y_{i+1} - \alpha_i y_{N+1}$. The contractivity of T on the complete metric space $(\mathcal{C}^1, \rho)$, yields a unique fixed point (say) f. Furthermore, to obtain the functional equation of the derivative f', define a complete metric space $(\mathcal{G}^*, \rho^*)$, where $\mathcal{G}^* = \{g^* \in \mathcal{C}(I, \mathbb{R}) | g^*(x_1) = d_1, g^*(x_{N+1}) = d_{N+1}\}$ and ρ^* is the uniform metric. Now the RB operator $\mathcal{T}^*$ on $(\mathcal{G}^*, \rho^*)$ is defined by

$$(\mathcal{T}^* g^*)(x) = \frac{\alpha_i g^*(L_i^{-1}(x)) + q_i'(L_i^{-1}(x))}{a_i}.$$

The unique fixed point of $\mathcal{T}^*$ is the required derivative f' satisfying

$$a_i f'(L_i(x)) = \alpha_i f'(x) + q_i'(x).$$

Here, the choice of cubic polynomial q_i should satisfy $f(x_1) = y_1$, $f(x_{N+1}) = y_{N+1}$, $f'(x_1) = d_1$, $f'(x_{N+1}) = d_{N+1}$. As the function f is defined piecewisely via the maps L_i, the cubic polynomial q_i is simply taken as,

$$q_i(x) = A_i \left(\frac{x - x_1}{x_{N+1} - x_1} \right)^3 + B_i \left(\frac{x - x_1}{x_{N+1} - x_1} \right)^2 + C_i \left(\frac{x - x_1}{x_{N+1} - x_1} \right) + D_i, \quad (3.3)$$

where the unknowns A_i, B_i, C_i, D_i are determined as follows

$$\begin{aligned}
A_i &= h_i(d_{i+1} + d_i) - \alpha_i(d_{N+1} + d_1)(x_{N+1} - x_1) - 2(y_{i+1} - y_i) + 2(y_{N+1} - y_1), \\
B_i &= -h_i(2d_i + d_{i+1}) + 3(y_{i+1} - y_i) - \alpha_i[-(x_{N+1} - x_1)(2d_1 + d_N) + 3(y_{N+1} - y_1)], \\
C_i &= h_i d_i - \alpha_i d_1(x_{N+1} - x_1), \\
D_i &= y_i - \alpha_i y_1,
\end{aligned}$$

here $h_i = x_{i+1} - x_i$. Taking $\theta = \frac{x - x_1}{x_{N+1} - x_1}$ and substituting the unknowns, it is seen that

$$\begin{aligned}
f(L_i(x)) = {} &\alpha_i f(x) + \big\{ h_i(d_{i+1} + d_i) - \alpha_i(d_{N+1} + d_1)(x_{N+1} - x_1) - 2(y_{i+1} - y_i) \\
&+ 2(y_{N+1} - y_1) \big\} \theta^3 + \big\{ -h_i(2d_i + d_{i+1}) + 3(y_{i+1} - y_i) \\
&- \alpha_i[-(x_{N+1} - x_1)(2d_1 + d_{N+1}) + 3(y_N - y_1)] \big\} \theta^2 \\
&+ \{ h_i d_i - \alpha_i d_1(x_{N+1} - x_1) \} \theta + y_i - \alpha_i y_1,
\end{aligned}$$

$$(3.4)$$

the function f satisfying Eq. (3.4) is called as the $\mathcal{C}^1$-cubic Hermite fractal interpolation function. The MATLAB code for generating Hermite fractal splines is provided as follows.

```
%A constructive approach to cubic Hermite fractal interpolation
    function
clc; clear all; close all;
x=[0  0.2  0.5  0.7  0.9];
y=[1  0.5  1.5  -1  2];
Data=[x' y']
%y=[5  4  8  6  7];
iter=8;
lx=length(x);
for i=1:lx-1
    h(i)=x(i+1)-x(i); hn=(x(lx)-x(1)); %Length of interval of first (
        n-1) values
    a(i)=h(i)/hn;
    b(i)=((x(i)*x(lx))-(x(i+1)*x(1)))/hn;
end
%d=Arithmetic_meansvalue(x,y)
alpha=[0.2  0.3  0.2  0.1]
%alpha=[0.1  0.3  0.1  0.2]
%alpha=[0  0  0  0]
d=cubicFIF_derivative(x,y,alpha)
[X Y]=Const_HermiteCubicFIF(x,y,alpha,d,iter);
plot(X,Y,'b-'); hold on
plot (x,y,'.k','markersize',20);
%%Generating new data points
function [X1 Y1]=Const_HermiteCubicFIF(x,y,alpha,d,iter)
lx=length(x);
for i=1:lx-1
    h(i)=x(i+1)-x(i); hn=(x(lx)-x(1)); %Length of interval of first (
        n-1) values.
    a(i)=h(i)/hn;
    b(i)=((x(i)*x(lx))-(x(i+1)*x(1)))/hn;
end
abvalue=[a' b']
N=length(x);
for i=1:lx-1
    %fprintf("---i value =%d",i)
    third=alpha(i)*[-hn*(d(1)+d(N))-2*(y(N)-y(1))];
    AA(i)=h(i)*(d(i)+d(i+1))-2*(y(i+1)-y(i))-alpha(i)*[hn*(d(1)+d(N
        ))-2*(y(N)-y(1))]; %
first term
    BB(i)=-h(i)*[2*d(i)+d(i+1)]+3*[y(i+1)-y(i)]-alpha(i)*[-hn*(2*d
        (1)+d(N))+3*(y(N)-
y(1))]; %second term
    CC(i)=h(i)*d(i)-alpha(i)*d(1)*hn; % third term
    DD(i)=y(i)-alpha(i)*y(1);
end
ABCD=[AA' BB' CC' DD'];
L=[]; L1=[]; X1=[]; Y1=[]; X=[]; Y=[];
p=lx;
for k=1:iter
    for i=1:lx-1
```

```
    for  t1 =1: p
      if  (k ==1)
        L(i , t1 )=(a(i )*x(t1 ))+b(i );
        theta (t1 )=(x(t1 )-x(1 ))/(x(p)-x(1 ));
        Q=(AA(i ))*(theta (t1 ))^3+BB(i )*(theta (t1 ))^2+CC(i )*(theta (
           t1 ))+DD(i );
        L1 (i , t1 )=(alpha (i )*y(t1 ))+Q;
      else
        %fprintf('----[i , t1 , p]=(%d,%d,%d)', i , t1 , p)
        L(i , t1 )=(a(i )*X1(t1 ))+b(i );
        theta (t1 )=(X1(t1 )-X1(1 ))/(X1(p)-X1(1 ));
        Q=(AA(i ))*(theta (t1 ))^3+BB(i )*(theta (t1 ))^2+CC(i )*(theta (
           t1 ))+DD(i );
        L1 (i , t1 )=(alpha (i )*Y1(t1 ))+Q;
      end
    end
    X=[X  L(i ,:) ];
    Y=[Y  L1 (i ,:) ];
  end
  X1=X;      Y1=Y;
  X =[];     Y =[];
  p=length (X1);
end
XX=[X1' Y1'];
XXX=unique (XX, 'rows');
X1=XXX(: ,1 ); Y1=XXX(: ,2 );
end
%%%%%%%%%%%%%%%%%%%%%%%%%%%%%%%%%
%% Finding the derivative values
function  d_new=cubicFIF_derivative (x ,y , alpha )
u1 =5.4523;  u2 =5.4523;
N=length (x );
for  n =1:N-1
  a(n)=(x(n+1)-x(n))/(x(N)-x(1 ));
  h(n)=x(n+1)-x(n);
end
% Computing co-efficient  matrix
A(1 )=((a(1 )* a(1 ))-alpha (1 ))*h(1 );%A1  Value
A1(1 )=4*a(1 )*a(1 )*(1 -((alpha (1 )/h(1 ))*(x(N)-x(1 ))));%A1*  Value .
mu (1 )= 2*a(1 )*a(1 );
B1(1 )=  -2*a(1 )*a(1 )*(alpha (1 )/h(1 ))*(x(N)-x(1 ));
beta (1 )=(6*a(1 )*a(1 )*(((y(2)-y(1 ))-(alpha (1 )*(y(N)-y(1 ))))/h(1 )))
   -(A(1 )*u1 );
B(N)=  -((a(N-1)*a(N-1))-alpha (N-1))*h(N-1);
B1(N)= 4*a(N-1)*a(N-1)*(1 -((alpha (N-1)/h(N-1))*(x(N)-x(1 ))));
A1(N)=-2* (alpha (N-1)/h(N-1))*a(N-1)*a(N-1)*(x(N)-x(1 ));
lam (N)=2*a(N-1)*a(N-1);
beta (N)=(6*a(N-1)*a(N-1)*(((y(N)-y(N-1))-(alpha (N-1)*(y(N)-y(1 )))
   )/h(N-1)))-
(B(N)* u2 );

% Computing the remaining values of co-efficient  matrix
for  n =2:N-1
  mu (n)= h(n-1)/(h(n-1)+h(n));
  lam (n)= 1-mu (n);
```

```
A(n) = -(alpha(n)*h(n)*h(n-1))/(2*a(n)*a(n)*(h(n)+h(n-1)));
B(n)= (alpha(n-1)*h(n)*h(n-1))/(2*a(n-1)*a(n-1)*(h(n)+h(n-1)));
A1(n)= -(1/(h(n)+h(n-1)))*(x(N)-x(1))*(2*alpha(n)*(h(n-1)/h(n))
    + alpha(n-1)*(h(n)/
h(n-1)));
B1(n)= -(1/(h(n)+h(n-1)))*(x(N)-x(1))*(alpha(n)*(h(n-1)/h(n))+
    2*alpha(n-1)*(h(n)/
h(n-1)));
beta(n)=(3*(h(n-1)/(h(n)+h(n-1)))* (((y(n+1)-y(n))- (alpha(n)*
    (y(N)-y(1)))))/
h(n)))+ (3*(h(n)/(h(n)+h(n-1)))* (((y(n)-y(n-1))- (alpha(n-1)*
    (y(N)-y(1)))))/
h(n-1)))-(A(n)*u1)-(B(n)*u2);
end
% Assign zero to the co-efficient matrix
C=zeros(N,N);
%The value of first column of The co-efficient matrix
for n=1:N
  if n ==2
    C(n,1)= A1(n) + lam(n);
  else
    C(n,1)= A1(n);
  end
end
% The value of last column of the co-efficient matrix
for n=1:N
  if n ==N-1
    C(n,N) = B1(n) + mu(n);
  else
    C(n,N) = B1(n);
  end
end
%The value of the intermediate columns

for n =2:N-1
  C(n-1,n)=mu(n-1);
  C(n,n)=2;
  C(n+1,n)=lam(n+1);
end
Cinv=inv(C);
% Take RHS in column vector
betat=beta ';
d = Cinv*betat;
deriv_value=d';
d_new=deriv_value;
end
%%%%%%%%%%%%%%%%%%%%%%%%%%%%%%%%%%%%%%%%
function [d]=Arithmetic_meansvalue(x,y)
N=length(x);
for n=1: N-1
  h(n)=x(n+1) - x(n);
  del(n) = (y(n+1)- y(n))/h(n);
end
ad(1) = del(1) + (h(1)*(del(1)-del(2)))/(h(1)+h(2));
ad(N) = del(N-1) + (h(N-1)*(del(N-1)-del(N-2)))/(h(N-1)+h(N-2));
```

```
gd(1) = del(1).^( 1+ (h(1)/h(2))) *((h(2)*del(2) + h(1)*del(1))/(h
    (1)+h(2))).^(-h(1)/
h(2)));
gd(N) = del(N-1).^(1+(h(N-1)/h(N-2))) *((h(N-2)*del(N-2) + h(N-1)*
    del(N-1))/(h(N-1)+
h(N-2))).^(-h(N-1)/h(N-2));
for n=2:N-1
    ad(n)= (h(n)*del(n-1)+ h(n-1)*del(n))/(h(n)+h(n-1));
    gd(n) = del(n-1).^( h(n)/(h(n-1)+h(n))) * del(n).^ (h(n-1)/(h(n
        -1)+h(n)));
end
d=ad;
end
```

3.1.1 Numerical Computation

Let $\{(0, 1), (0.2, 0.5), (0.5, 1.5)(0.7, -1), (0.9, 2)\}$ be the given set of interpolation data. For the construction of Hermite cubic FIF, the derivatives are computed using the reference [2]. To observe the effect of scaling parameters, different choice of scaling constants are chosen as given in Table 3.1. The different Hermite cubic FIFs are shown in Fig. 3.1a, b and c by modifying α_i and derivative values with the help of arithmetic mean method. The classical Hermite cubic FIF is constructed with the zero scaling vector and its graph is shown in Fig. 3.1d.

3.2 Cubic Fractal Spline Using Moments

In [3], moments are used to construct the fractal splines. The moments M_i are defined by

$$M_i = f''(x_i), \quad i = 1, 2, \ldots, N + 1.$$

Suppose the function $f \in C^2[x_1, x_N]$ and whose graph is the fixed point of the iterated function system $\{(L_i(x), F_i(x, y)), i = 1, 2, \ldots, N - 1\}$ satisfying $f(x_i) = y_i$, where

Table 3.1 Scaling parameters and derivatives associated with Hermite cubic FIF

Scaling parameters	Derivatives
[0.2, 0.3, 0.2, 0.2]	[36.8326, 69.8087, 45.9254, 54.3726, 95.5692]
[0.1, 0.3, 0.1, 0.2]	[−0.2222, 45.7367, 34.8190, 16.1667, 119.8519]
[0.2, 0.3, 0.2, 0.1]	[−2.6762, 18.7817, 0.15568.6348, 30.0928]
[0, 0, 0]	[−5.6932, 3.3412, −9.4163, −1.6732, 23.6092]

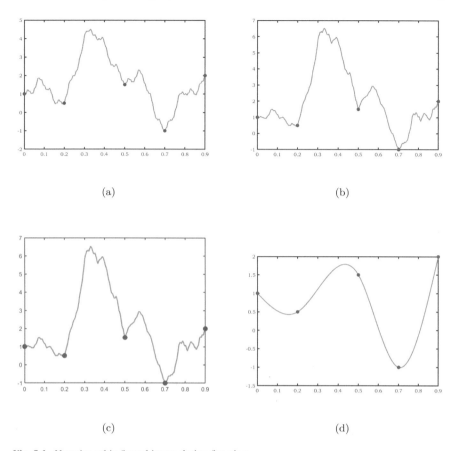

Fig. 3.1 Hermite cubic fractal interpolation function

$$L_i(x) = a_i x + b_i,$$
$$F_i(x, y) = a_i^2 \alpha_i y + a_i^2 q_i(x),$$

with $0 < |\alpha_i| < 1$ and q_i being a cubic polynomial, then f is said to be cubic spline fractal interpolation function. From the narration of spline FIF in the above sections, it follows that

$$f''(L_i(x)) = \alpha_i f''(x) + \frac{c_i(x - x_1)}{x_N - x_1} + d_i. \tag{3.5}$$

From Eqs. (2.3), (3.5) and the assumption $M_i = f''(x_i)$, the parameters c_i and d_i are determined as

$$c_i = M_{i+1} - M_i - \alpha_i(M_N - M_1),$$
$$d_i = M_i - \alpha_i M_1.$$

Then, for $i = 1, 2, \ldots, N$,

$$f''(L_i(x)) = \alpha_i f''(x) + \frac{(M_{i+1} - \alpha_i M_{N+1})(x - x_1)}{x_{N+1} - x_1} + \frac{(M_i - \alpha_i M_1)(x_{N+1} - x)}{x_{N+1} - x_1}.$$

On integrating the function f'' twice,

$$f(L_i(x)) = a_i^2 \{\alpha_i f(x) + \frac{(M_{i+1} - \alpha_i M_{N+1})(x - x_1)^3}{6(x_{N+1} - x_1)} + \frac{(M_i - \alpha_i M_1)(x_{N+1} - x)^3}{6(x_{N+1} - x_1)}$$
$$+ c_i^*(x_i - x) + d_i^*(x - x_1)\},$$

$$(3.6)$$

With $L_i(x_1) = x_i$, $L_i(x_{N+1}) = x_{i+1}$ and the interpolation conditions, the constants c_i^* and d_i^* are obtained as

$$c_i^* = \frac{1}{(x_{N+1} - x_1)}\left(\frac{y_i}{a_i^2} - \alpha_i y_1\right) - \frac{(M_i - \alpha_i M_1)(x_{N+1} - x_1)}{6},$$
$$d_i^* = \frac{1}{(x_{N+1} - x_1)}\left(\frac{y_{i+1}}{a_i^2} - \alpha_i y_{N+1}\right) - \frac{(M_{i+1} - \alpha_i M_{N+1})(x_{N+1} - x_1)}{6}.$$

On substituting c_i^* and d_i^* in the Eq. (3.6), the functional equation of f can be obtained in terms of moments. The following is the MATLAB code corresponding to the above discussed theory for developing fractal functions using moments.

```
%Fitting the  cubic spline FIF using derivative boundary
     conditions y'(x_1)=fd1 and y'(x_N)=fd2.
function []= Cubic_itperpolation ()
clc ; clear  all ; close  all ;
x =[50  60  72  100]; y =[82  50  78  40  ];%Data set
n=length (x) ;
iter =input ('Enter the  no.of  interations :=');
Size =n
Alpha =input ('Enter the  alpha   values (Size -1):=');
m=Moments (x ,y , Alpha ) ;%Finding  the  moments
[X,Y]= Cubic_simplification (x ,y , Alpha , iter ,m) ;
XYvalues =[X  Y];
%% Ploting  graph
plot (x ,y ,'.k','markersize ',30); hold  on ; plot (X,Y,'.b','markersize
     ',4); hold  on ; plot (X,Y,'r-') ;
end
%% Generated  data (Using  Moments)
function [X1  Y1]= Cubic_simplification (x ,y , alpha , iter ,m)

n=length (x) ;
p=n ;
for  i =1:n-1
  a (i) =(x (i+1)-x (i))/(x (n)-x (1)) ;
  b (i) =((x (n)*x (i)) -(x (1)*x (i+1)))/(x (n)-x (1)) ; a2 (i)=b (i) ;
end
abvalues =[a'  b']
```

```
for  i =1:n−1
   q =(x(n)−x(1));
   q1=m(i+1)−alpha(i)*m(n);
   q2=m(i)−alpha(i)*m(1);
   q3=(y(i)/(a(i)^2))−alpha(i)*y(1);
   q4=(y(i+1)/a(i)^2)−alpha(i)*y(n);
   s1(i)=power(a(i),2)*(q1−q2)/(6*q);s2(i)=power(a(i),2)*(x(n)*q2−
      x(1)*q1)/(2*q);
   s3(i)=power(a(i),2)*(((3*q1*power(x(1),2))−(3*q2*power(x(n),2))
      )/(6*q)+(q2−q1)*(q/6)+(q4−q3)/q);
   s4(i)=power(a(i),2)*(((q2*power(x(n),3))−(q1*power(x(1),3)))
      /(6*q)+(x(1)*q1−x(n)*q2)*(q/6)+(x(n)*q3−x(1)*q4)/q);
   s5(i)=power(a(i),2)*alpha(i);
end
%%%%%
%%%
X=[];Y=[];
for  k=1:iter
   for  i =1:n−1
      for  t =1:p
         if  (k==1)
            L(i,t)=(a(i)*x(t))+b(i);
            L1(i,t)=s5(i)*y(t)+s1(i)*power(x(t),3)+s2(i)*power(x(t)
               ,2)+s3(i)*x(t)+s4(i);

         else
            L(i,t)=(a(i)*X1(t))+b(i);
            L1(i,t)=s5(i)*Y1(t)+s1(i)*power(X1(t),3)+s2(i)*power(X1(t
               ),2)+s3(i)*X1(t)+s4(i);

         end
      end
      X=[X L(i,:)];
      Y=[Y L1(i,:)];
   end
   g =[X' Y'];
   g=str2num(num2str(g,10));g=unique(g,'rows');
   X1=g(:,1); Y1=g(:,2); p=length(X1);X=[];Y=[];
end
end
%%Finding the moments
function   [m]=Moments(x,y,alpha)
n=length(x);
for  i =1:n−1% Here finding a,b,c,d values
   a(i)=(x(i+1)−x(i))/(x(n)−x(1));
   b(i)=((x(n)*x(i))−(x(1)*x(i+1)))/(x(n)−x(1));
end
for  i =1:n−1
   h(i)=x(i+1)−x(i);
end
for  i =1:n
   if  i ==1
      as(i)=6*(1−a(i)*alpha(i));ca(i)=2*((1−alpha(i))*h(i));la(i)=h
         (i);cb(i)=−alpha(i)*h(i);
      d(i)=(y(i+1)−y(i)−(alpha(i)*(a(i)^2)*(y(n)−y(i))))*(6/h(i));
```

```
elseif i==n
    ca(i)=-alpha(i-1)*h(i-1); mu(i)=h(i-1);cb(i)=2*(1-alpha(i-1))
        *h(i-1);bs(i)=-6*(1-a(i-1)*alpha(i-1));
    d(i)=-(y(i)-y(i-1)-(alpha(i-1)*(a(i-1)^2)*(y(i)-y(1))))*(6/h(
        i-1));
else
    as(i)=-6*((a(i)*alpha(i))/(h(i-1)+h(i)));
    ca(i)=-((alpha(i-1)*h(i-1)+2*(alpha(i)*h(i))))/(h(i)+h(i-1));
    la(i)=h(i)/(h(i-1)+h(i));mu(i)=1-la(i);cb(i)=-(((2*alpha(i-1)
        *h(i-1))+(alpha(i)*h(i)))/(h(i-1)+h(i)));
    bs(i)=6*(a(i-1)*alpha(i-1)/(h(i-1)+h(i)));
    d1(i)=(((y(i+1)-y(i))/h(i)) -(((y(i)-y(i-1))/h(i-1))));
    d2(i)=((a(i)*alpha(i))-(a(i-1)*alpha(i-1)))*(y(n)-y(1))/(x(n)
        -x(1)));
    d(i)=6*(d1(i)-d2(i))/(h(i)+h(i-1))  ;
end
end
fd1=input('Enter the Initial Derivative value:=');
fd3=input('Enter the End Derivative value:=');
fd1=2;fd3=5;% Here using derivative boundary condition of type 1.
dd=[d(1)-as(1)*fd1   d(2)-as(2)*fd1-bs(2)*fd3  d(3)-as(3)*fd1-bs(3)
    *fd3 d(4)-bs(4)*fd3];
aa=[ca(1) la(1) 0 cb(1);ca(2)+mu(2) 2 la(2) cb(2);ca(3) mu(3) 2
    la(3)+cb(3);ca(4) 0 mu(4) cb(4)];
m=inv(aa)*dd';
m=m';
Moments_values=m
end
```

3.2.1   Numerical Computation

Let $\{(50, 82), (51, 50), (52, 78), (53, 40)\}$ be the given data set. With the assumptions $f'(x_1) = 100$ and $f'(x_N) = 10$, the C^2-cubic spline FIFs are computed. The system of equations is solved using the reference [3]. Different set of scaling factors and moments are provided in Table 3.2. The effects of perturbation in the scaling factor α with respect to the IFS parameters are shown in Fig. 3.2. The graphs of generated C^2 continuity of the cubic FIF are shown in Fig. 3.2a and b with the modified scalings. Figure 3.2c demonstrates the classical cubic FIF, which is retrieved (from Fig. 3.2a) by setting all the scaling factors to be zero.

Table 3.2 Scaling parameters and moments associated with the C^2-cubic spline FIF

Scaling parameters	Moments
[0.8, 0.8, 0.9]	[−214.6452, 433.1183, −100.7957, 657.0753]
[0.9, 0.9, 0]	[−416.1359, 106.7961, −183.7087, 220.8544]
[0, 0, 0]	[−2.3245, 1.5290, −0.9025, 1.1323]

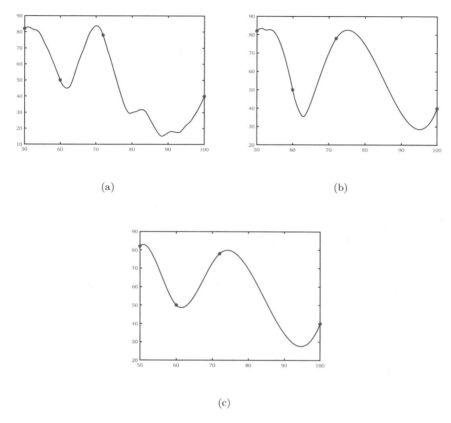

Fig. 3.2 Cubic fractal interpolation function with moments

3.3   Rational Fractal Spline

The rational cubic fractal interpolation function with numerator as cubic polynomial and denominator as linear function is defined in [4] by

$$f(L_i(x)) = \alpha_i f_i + \frac{P_i(\theta)}{Q_i(\theta)}, \tag{3.7}$$

where,

$$
\begin{aligned}
P_i(\theta) &= r_i(y_i - \alpha_1 y_1)(1 - \theta)^3 + t_i(y_{i+1} - \alpha_1 y_N)\theta^3 + \big\{(2r_i + t_i)y_i + r_i h_i d_i \\
&\quad - \alpha_i[(2r_i + t_i)y_1 + r_i(x_N - x_1)]\big\} + \big\{(r_i + 2t_i)y_{i+1} - t_i h_i d_{i+1} \\
&\quad - \alpha_i[(r_i + t_i)y_N - t_i(x_N - x_1)]\big\} \\
Q_i(\theta) &= r_i(1 - \theta)r_i + t_i\theta.
\end{aligned}
$$

Consider the prescribed data set $\{(x_i, y_i) : i = 1, 2, \ldots, N + 1\}$. The derivatives at the knots x_i are denoted by d_i. For simplicity, take $h_i = x_{i+1} - x_i$, for $i = 1, 2, \ldots, N - 1$. Let r_i and t_i be the free shape parameters, in order to maintain the positiveness of the denominator in rational fractal splines, the parameters are restricted to be $r_i > 0$ and $t_i > 0$. Readers are recommended to see the articles [5, 6] for the construction of rational cubic spline fractal functions. The following equation provides the C^1 rational cubic spline with numerator as cubic polynomial and the denominator as linear function,

$$
f(L_i(x)) = \frac{(1 - \theta)^3 r_i y_i + \theta(1 - \theta)^2 V_i + \theta^2(1 - \theta)W_i + \theta^3 t_i y_{i+1}}{(1 - \theta)r_i + \theta t_i},
$$

here,

$$
\begin{aligned}
V_i &= (2r_i + t_i)y_i + r_i h_i d_i, \\
W_i &= (r_i + 2t_i)y_{i+1} - t_i h_i d_{i+1}, \\
\theta &= \frac{x - x_1}{x_{N+1} - x_1}.
\end{aligned}
$$

To construct the fractal perturbation f^α of the rational cubic spline , the scaling factor is chosen so that $|\alpha_i| < a_i$ and the family of base functions are defined by

$$
b_i(x) = \frac{B_{1i}(1 - \theta)^3 + B_{2i}\theta(1 - \theta)^2 + B_{3i}\theta^2(1 - \theta) + B_{4i}\theta^3}{(1 - \theta)r_i + \theta t_i},
$$

such that each function b_i should agree the prescribed function f at the end points of the interval of interpolation. The coefficients are given by

$$
\begin{aligned}
B_{1i} &= r_i y_1, \\
B_{2i} &= (2r_i + t_i)y_1 + r_i d_1(x_{N+1} - x_1), \\
B_{3i} &= (r_i + 2t_i)y_{N+1} - t_i d_{N+1}(x_{N+1} - x_1), \\
B_{4i} &= t_i y_{N+1}.
\end{aligned}
$$

Now the C^1 rational cubic fractal spline is expressed by

$$f^\alpha(L_i(x)) = \alpha_i f^\alpha(x) + \frac{P_i(x)}{Q_i(x)}, \tag{3.8}$$

where,

$$
\begin{aligned}
P_i(x) = P_i^*(\theta) &= (y_i - \alpha_i y_1)r_i(1-\theta)^3 + (y_{i+1} - \alpha_i y_{N+1})t_i\theta^3 + \{(2r_i + t_i)y_i + r_i h_i d_i \\
&\quad - \alpha_i[(2r_i + t_i)y_1 + r_i(x_N - x_1)d_1]\}\theta(1-\theta)^2 + \{(r_i + 2t_i)y_{i+1} - t_i h_i d_{i+1} \\
&\quad - \alpha_i[(r_i + 2t_i)y_{N+1} - t_i(x_{N+1} - x_1)d_{N+1}]\}\theta^2(1-\theta), \\
Q_i(x) = Q^*(\theta) &= (1-\theta)r_i + \theta t_i, \quad i = 1, 2, \ldots, N, \\
\theta &= \frac{x - x_1}{x_{N+1} - x_1}.
\end{aligned}
$$

The MATLAB code is provided to produce the rational cubic fractal functions.

```
%Rational  FIF  with  cubic  as  numerator  and  denominator  as  linear
clc;clear  all;close  all;
x=[0  2  3  6  7];y=[5    4    8    6  7];
iter=6;
lx=length(x);
for  i=1:lx-1
  h(i)=x(i+1)-x(i);hn=(x(lx)-x(1));%Length  of  interval  of  first  (
      n-1)  values.
  a(i)=h(i)/hn;
  b(i)=((x(i)*x(lx))-(x(i+1)*x(1)))/hn;
end
[a  b]
%r=1000*[3.1  1  1  1];t=[1  1  1  1];alpha=[0.24  0.1  0.35  0.1]%%
      Example1
%r=[3.1  1  1  1];t=[1  1  1  1];alpha=[0.24  0.1  0.35  0.1]% Example2
r=1000*[3.1  1  1  1];t=[1  1  1  1];alpha=[  0.1  0.1  0.1  0.1]%Example3
%r=1000*[3.1  1  1  1];t=[1  1  1  1];alpha=[0  0  0  0]% Classical
d=Arthemetic_meansvalue(x,y)
[X  Y]=Const_CubicFIF(x,y,r,t,alpha,d,iter);
plot(X,Y,'b-');hold  on
plot  (x,y,'.k','markersize',20);
%%%%%%%%%%%%%%%%%
function  [X1  Y1]=Const_CubicFIF(x,y,r,t,alpha,d,iter)
lx=length(x);
for  i=1:lx-1
  h(i)=x(i+1)-x(i);hn=(x(lx)-x(1));%Length  of  interval  of  first  (
      n-1)  values.
  a(i)=h(i)/hn;
  b(i)=((x(i)*x(lx))-(x(i+1)*x(1)))/hn;
end
avalue=a
for  i=1:lx-1
  cf1(i)=(y(i)-alpha(i)*y(1))*r(i);%first  term
  cf2(i)=(y(i+1)-alpha(i)*y(lx))*t(i);  %  secondterm
  cf31(i)=(2*r(i)+t(i))*y(i)+r(i)*h(i)*d(i);
  cf32(i)=(2*r(i)+t(i))*y(1)+r(i)*hn*d(1);
```

```
      cf3 ( i )=cf31 ( i )−alpha ( i )∗cf32 ( i ) ;
      cf41 ( i )=( r ( i )+2∗t ( i ) )∗y ( i +1)−t ( i )∗h ( i )∗d ( i +1) ;
      cf42 ( i )=( r ( i )+2∗t ( i ) )∗y ( lx )−t ( i )∗hn∗d ( lx ) ;
      cf4 ( i )=cf41 ( i )−alpha ( i )∗cf42 ( i ) ;
   end
%[ a ' b ' cf1 ' cf2 ' cf3 ' cf4 ' ]

   L =[ ] ; L1 =[ ] ;
   X1 =[ ] ; Y1 =[ ] ;
   X =[ ] ; Y =[ ] ;
   p=lx ;
   for  k=1: iter
      for  i =1: lx −1
         for  t1 =1: p
            if  ( k ==1)
               L( i , t1 )=( a ( i )∗x ( t1 ) )+b ( i ) ;
               theta ( t1 )=( x ( t1 )−x ( 1 ) )/( x ( p )−x ( 1 ) ) ;
               L11 =( cf1 ( i ) )∗(1− theta ( t1 ) )^3;
               L12 =( cf2 ( i ) )∗( theta ( t1 ) )^3;
               L13=cf3 ( i )∗( theta ( t1 )∗(1− theta ( t1 ) )^2) ;
               L14=cf4 ( i )∗(( theta ( t1 ) )^2∗(1− theta ( t1 ) ) ) ;
               px=L11+L12+L13+L14 ;
               pxx=(1− theta ( t1 ) )∗r ( i )+theta ( t1 )∗t ( i ) ;
               L1( i , t1 )=( alpha ( i )∗y ( t1 ) )+( px / pxx ) ;
            else
               L( i , t1 )=( a ( i )∗X1( t1 ) )+b ( i ) ;
               theta ( t1 )=( X1( t1 )−X1( 1 ) )/( X1( p )−X1( 1 ) ) ;
               L11 =( cf1 ( i ) )∗(1− theta ( t1 ) )^3;
               L12 =( cf2 ( i ) )∗( theta ( t1 ) )^3;
               L13=cf3 ( i )∗( theta ( t1 )∗(1− theta ( t1 ) )^2) ;
               L14=cf4 ( i )∗(( theta ( t1 ) )^2∗(1− theta ( t1 ) ) ) ;
               px=L11+L12+L13+L14 ;
               pxx=(1− theta ( t1 ) )∗r ( i )+theta ( t1 )∗t ( i ) ;
               L1( i , t1 )=( alpha ( i )∗Y1( t1 ) )+( px / pxx ) ;
            end
         end
         X=[X  L( i ,:) ] ;
         Y=[Y  L1( i ,:) ] ;
      end
      X1=X ;        Y1=Y ;
      X =[ ] ;        Y =[ ] ;
      p=length ( X1 ) ;
   end
   XX=[X1 '  Y1 ' ] ;
   XXX=unique ( XX , 'rows ' ) ;
   X1=XXX( : , 1 ) ; Y1=XXX( : , 2 ) ;
   end

   function  [ d ]= Arithmetic_meansvalue ( x , y )
   N=length ( x ) ;
   for  n =1:  N−1
      h ( n )=x ( n+1) −  x ( n ) ;
      del ( n )  =  ( y ( n+1)−  y ( n ) )/ h ( n ) ;
   end
   ad ( 1 )  =  del ( 1 )  +  ( h ( 1 )∗( del ( 1 )−del ( 2 ) ) )/( h ( 1 )+h ( 2 ) ) ;
```

```
ad(N) = del(N-1) + (h(N-1)*(del(N-1)-del(N-2)))/(h(N-1)+h(N-2));
gd(1) = del(1).^( 1+ (h(1)/h(2)))*((h(2)*del(2) + h(1)*del(1))/(h
    (1)+h(2))).^(-h(1)/
h(2));
gd(N) = del(N-1).^(1+(h(N-1)/h(N-2)))*((h(N-2)*del(N-2) + h(N-1)*
    del(N-1))/(h(N-1)+
h(N-2))).^(-h(N-1)/h(N-2));
for n=2:N-1
  ad(n)= (h(n)*del(n-1)+ h(n-1)*del(n))/(h(n)+h(n-1));
  gd(n) = del(n-1).^( h(n)/(h(n-1)+h(n))) * del(n).^ (h(n-1)/(h(n
    -1)+h(n)));
end
d=ad;
end
```

3.3.1 Numerical Computation

Consider the data set $\{(0, 5), (2, 4), (3, 8), (6, 6), (7, 7)\}$. The derivative values (d_i, $i = 1, 2, 3, 4, 5$) are estimated using the arithmetic mean method: $d_1 = -3.5$, $d_2 = 2.5$, $d_3 = 2.8333$, $d_4 = 0.5833$, $d_5 = 1.4167$. The rational cubic FIFs are generated with the scaling parameters and shape parameters as given in Table 3.3. The graphs of rational cubic FIFs are illustrated in Fig. 3.3a, b and c. The classical version of rational cubic FIF is obtained with the choice $\alpha = (0, 0, 0, 0)$ and its corresponding graph is demonstrated in Fig. 3.3d.

Table 3.3 Scaling parameters and shape parameters associated with the rational FIF

Scaling parameters	Shape parameters	Figure
[0.24, 0.1, 0.35, 0.1]	$r = 1000 * [3.1, 1, 1, 1]$, $t = [1, 1, 1, 1]$	Figure 3.3a
[0.24, 0.1, 0.35, 0.1]	$r = [3.1, 1, 1, 1]$, $t = [1, 1, 1, 1]$	Figure 3.3b
[0.1, 0.1, 0.1, 0.1]	$r = 1000 * [3.1, 1, 1, 1]$, $t = [1, 1, 1, 1]$	Figure 3.3c
[0, 0, 0, 0]	$r = 1000 * [3.1, 1, 1, 1]$, $t = [1, 1, 1, 1]$	Figure 3.3d

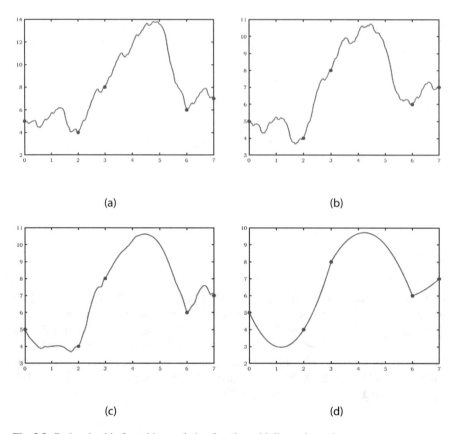

(a)   (b)

(c)   (d)

Fig. 3.3 Rational cubic fractal interpolation function with linear denominator

References

1. M.F. Barnsley, A.N. Harrington, The calculus of fractal interpolation functions. J. Approx. Theory **57**(1), 14–34 (1989)
2. A.K.B. Chand, P. Viswanathan, A constructive approach to cubic Hermite fractal interpolation function and its constrained aspects. BIT **53**(4), 841–865 (2013)
3. A.K.B. Chand, G.P. Kapoor, Generalized cubic spline fractal interpolation functions. SIAM J. Numer. Anal. **44**(2), 655–676 (2006)
4. A.K.B. Chand, P. Viswanathan, K.M. Reddy, Towards a more general type of univariate constrained interpolation with fractal splines. Fractals **23**(4), 1550040, 12 (2015)
5. S.K. Katiyar, A.K.B. Chand, G. Saravana Kumar, A new class of rational cubic spline fractal interpolation function and its constrained aspects. Appl. Math. Comput. **346**, 319–335 (2019)
6. N. Balasubramani, M. Guru Prem Prasad, S. Natesan, Shape preserving α-fractal rational cubic splines. Calcolo **57**(3), 21 (2020)

Chapter 4
Fractal Interpolation Surfaces

The most general form of data encountered in real life problems is three dimensional data that can be visualized as surfaces. Surface interpolates play vital role in industry, geology, diagnosis and CAD [1]. The usual solution to a problem of surface interpolation is to determine a bivariate function $z = f(x, y)$, which assumes finite discrete values in a given domain. However, the construction of surface interpolation function is not an easy process comparing to univariate classical interpolations. For more details on the construction of fractal interpolation surfaces, visit [2–5].

4.1 Construction of Fractal Surfaces

First we will construct Rational Fractal Interpolation Function (RCFIFs) along the grid lines in the domain of surface interpolation data in Sect. 4.1.1. In Sect. 4.1.2, a partially blended RCFIF is constructed by using the univariate RCFIF and blending functions.

4.1.1 RCFIFs Along X-direction and Y-direction

Consider a surface interpolation data set $\{(x_i, y_j, f_{i,j}, f^x_{i,j}, f^y_{i.j}) : i \in N_m, j \in N_n\}$, where N_n denotes first n-natural numbers.

RCFIFs along X-direction: For each $j \in N_n$ (along the j-th grid line parallel to x-axis), the construction of univariate FIFs $\Psi(x, y_j)$ and $\Psi(x, y_{j+1})$ are presented. Consider

$$\Psi(x, y_j) = \alpha_{i,j}(L_i^{-1}(x), y_j) + \frac{P_{i,j}(\theta)}{Q_{i,j}(\theta)}, \quad i \in \mathbb{N}_M, \tag{4.1}$$

© The Author(s), under exclusive license to Springer Nature Switzerland AG 2023
S. Banerjee et al., *Fractal Patterns with MATLAB*,
SpringerBriefs in Complexity,
https://doi.org/10.1007/978-3-031-48102-4_4

where,

$$P_{i,j}(\theta) = (f_{i,j} - \alpha_{i,j} f_{1,j}(1-\theta)^3 + \{r_{i,j} f_{i,j} + h_i f_{i,j}^x - \alpha_i [r_{i,j} f_{1,j}$$
$$+ (x_M - x_1) f_{1,j}^x\}(1-\theta)^2\theta + \{r_{i,j} f_{i+1,j} - h_i f_{i+1,j}^x - \alpha_i [r_{i,j} f_{M,j}$$
$$+ (x_M - x_1) f_{M,j}^X\}(1-\theta)\theta^2 + f_{i,j} - \alpha_{i,j} f_{1,j}(1-\theta)^3,$$

$$Q_{i,j} = 1 + (r_{i,j} - 3)\theta(1-\theta), \quad \theta = \frac{x - x_i}{h_i}, \quad x \in I_i.$$

RCFIFs along Y-direction: For each $i \in N_m$ (along the i-th grid line parallel to y-axis), rational cubic spline fractal interpolation functions $\Psi^*(x_i, y)$ and $\Psi^*(x_{i+1}, y)$ are presented. Consider

$$\Psi^*(x_i, y) = \alpha_{i,j}^*(x_i, L^{*-1}(y)) + \frac{P_{i,j}^*(\phi)}{Q_{i,j}^*(\phi)}, \quad j \in \mathbb{N}_N, \qquad (4.2)$$

where,

$$P_{i,j}^*(\phi) = (f_{i,j} - \alpha_{i,j}^* f_{1,j}(1-\phi)^3 + \{r_{i,j} f_{i,j} + h_j^* f_{i,j}^x - \alpha_i^* [r_{i,j} f_{1,j}$$
$$+ (x_M - x_1) f_{1,j}^x\}(1-\phi)^2\phi + \{r_{i,j} f_{i+1,j} - h_j^* f_{i+1,j}^x - \phi_i^* [r_{i,j} f_{M,j}$$
$$+ (x_M - x_1) f_{M,j}^X\}(1-\phi)\phi^2 + f_{i,j} - \alpha_{i,j}^* f_{1,j}(1-\phi)^3,$$

$$Q_{i,j}^*(\phi) = 1 + (r_{i,j} - 3)\phi(1-\phi), \quad \phi = \frac{y - y_j}{h_j^*}, \quad y \in J_j.$$

4.1.2   Fractal Surfaces by Coon's Technique

Let us consider a Hermite surface interpolation data set $\{(x_i, y_j, f_{i,j}, f_{i,j}^x, f_{i,j}^y) : i \in N_m, \; j \in N_n\}$. The construction of univariate FIFs $\Psi(x, y_j)$, $\Psi(x, y_{j+1})$. And $\Psi^*(x_i, y)$, $\Psi^*(x_{i+1}, y)$ are discussed with cubic Hermite functions in the following. Let

$$b_{0,3}^i(x) = (1-\theta)^2(1+2\theta),$$
$$b_{3,3}^i(x) = \theta^2(3-2\theta),$$
$$b_{0,3}^j(y) = (1-\phi)^2(1+2\phi),$$
$$b_{3,3}^j(y) = \phi^2(3-2\phi).$$

Fig. 4.1 Continuity domain

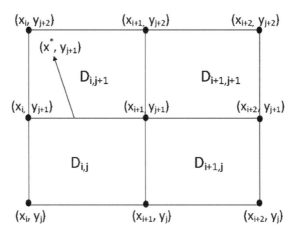

On each individual patch $D_{i,j} = I_i \times J_j$, $i \in N_{m-1}$, $j \in N_{n-1}$ (see Fig. 4.1), a blending rational cubic spline FIS is defined using the blending coons technique as

$$\Phi(x, y) = -\left[-1 \; b_{0,3}^i(x) \; b_{3,3}^i(x) \right] \begin{bmatrix} 0 & \Psi(x, y_j) & \Psi(x, y_{j+1}) \\ \Psi^*(x_i, y) & f_{i,j} & f_{i,j+1} \\ \Psi^*(x_{i+1}, y) & f_{i+1,j} & f_{i+1,j+1} \end{bmatrix} \begin{bmatrix} -1 \\ b_{0,3}^j(y) \\ b_{3,3}^j(y) \end{bmatrix}.$$

The function interpolation surface Φ interpolate the given data at grid points. The MATLAB code of rational cubic FIS [6, 7] is illustrated in the following.

```
%Fractal surface R(X)=P(X)/Q(X) from  P(X) is cubic and  Q(X) is
     Quadratic.
%Here scaling factor is alpha and  shape parameter is r.
%Inputs   are (x,y,z,alphax,alphay,rx,ry,partial derivatives(dx,
     dy) ).
%Outputs are "fractal surfaces".
clc;clear all;close all;
format('short')
x=[0 2 6 11];y=[0 3 7 8];%Given Data
iter=4;
grid on
m=length(x);n=length(y);
z=[1 10 8 11 ;2 11 9 12;3 12 10 14;4 13 12 15];
lx=length(x);ly=length(y);N=length(x);M=length(y);
p1=lx;p2=ly;
dx=[4.5 1.5 0.13 1.37;4.5 1.5 0.12 1.37;4.5 1.5 0.25 1.75;4.45
     1.61 0.25 1.25];
%dx=dx'
dy=[5 2.5 0.42 0.1;0.5 0.5 0.5 0.1;0.5 0.5 0.5 0.5;0.25 0.75 0.75
     0.25];
```

```
%dy=dy'
% Free Shape parameters
rx =0.5*ones(lx −1,ly);
ry =100*ones(lx ,ly −1);
alphax =[0.2  0.25  0.25  0.2;0.35  0.3  0.35  0.34;0.35  0.35  0.35
    0.34];
alphay =[0.124  0.562  0.30;0.125  0.562  0.30;0.125  0.562  0.3;0.125
    0.562  0.3];
for  i =1:lx −1
  h(i)=x(i +1)−x(i);
end
for  l =1:ly −1
  hy(l)=y(l +1)−y(l);
end
%***** Definitions of Matrices*****
a=zeros(1,lx −1);b=zeros(1,lx −1);
Lx =[];Lx1 =[];Lx2 =[];
X_na =[];X1 =[];Zx1 =[];
X =[];Zx2 =[];
a=zeros(1,ly −1);b=zeros(1,ly −1);
Lyy =[];Ly1 =[];Ly2 =[];
Y1 =[];Y =[];
Zy1 =[];Zy2 =[];
L =[];L1 =[];L2 =[];
L22 =[];L222 =[];
X11 =[];Y11 =[];
Z11 =[];Z22 =[];
XX =[];YY =[];ZZ =[];
N1=N;
M1=M;
%%Finding co−efficients
for  k =1:iter
  %*****X−direction(Y is fixed)*****
  for  j =1:3
    for  i =1:lx −1
      %%——Finding co−efficients if Fixing f(x,y(j))
      a(i)=(x(i +1)−x(i))/(x(lx)−x(1));
      b(i)=((x(i)*x(lx))−(x(i +1)*x(1)))/(x(lx)−x(1));
      Ax(i ,j)=(z(i ,j)−alphax(i ,j)*z(1,j))*rx(i ,j);
      Dx(i ,j)=(z(i +1,j)−(alphax(i ,j)*z(lx ,j)))*rx(i ,j);
      Bx1(i ,j)=rx(i ,j)*z(i ,j)+h(i)*dx(i ,j);
      Bx2(i ,j)=rx(i ,j)*z(1,j)+(x(lx)−x(1))*dx(1,j);
      Bx(i ,j)=Bx1(i ,j)−alphax(i ,j)*Bx2(i ,j);
      Cx1(i ,j)=rx(i ,j)*z(i +1,j)−h(i)*dx(i +1,j);
      Cx2(i ,j)=rx(i ,j)*z(lx ,j)−(x(lx)−x(1))*dx(lx ,j);
      Cx(i ,j)=Cx1(i ,j)−alphax(i ,j)*Cx2(i ,j);
      %%——Finding co−efficients if Fixing f(x, y(j +1))
      Axx(i ,j)=(z(i ,j +1)−(alphax(i ,j +1)*z(1,j +1)))*rx(i ,j +1);
      Dxx(i ,j)=(z(i +1,j +1)−(alphax(i ,j +1)*z(lx ,j +1)))*rx(i ,j +1);
      Bxx1(i ,j)=rx(i ,j +1)*z(i ,j +1)+h(i)*dx(i ,j +1);
      Bxx2(i ,j)=rx(i ,j +1)*z(1,j +1)+(x(lx)−x(1))*dx(1,j +1);
      Bxx(i ,j)=Bxx1(i ,j)−alphax(i ,j +1)*Bxx2(i ,j);
      Cxx1(i ,j)=rx(i ,j +1)*z(i +1,j +1)−h(i)*dx(i +1,j +1);
      Cxx2(i ,j)=rx(i ,j)*z(lx ,j)+(x(lx)−x(1))*dx(lx ,j +1);
      Cxx(i ,j)=Cxx1(i ,j)−alphax(i ,j +1)*Cxx2(i ,j);
```

```
end% end for loop i
abvalues =[a' b']
    theta=inline('(x-0)/11');
for i=1:lx-1
  for t1 =1:p1
    if (k==1)
      Lx(i,t1)=(a(1,i)*x(1,t1))+b(1,i);
      L11(i,t1)=t1+(i-1)*p1;
      %Fixing f(x, y(j))
      Qx=1+(rx(i,j)-3)*(1-theta(x(t1)))*(theta(x(t1)));
      Lx1(i,t1)=(alphax(i,j)*z(t1,j))+(((Ax(i,j)*(1-theta(x(
          t1)))^3)+(Bx(i,j)*theta(x(t1))*(1-theta(x(t1)))^2)
          +...
        (Cx(i,j)*(theta(x(t1))^2)*(1-theta(x(t1))))+(Dx(i,j)
          *(theta(x(t1)))^3))/Qx);
      %Fixing f(x, y(j+1))
      Qxx=1+(rx(i,j+1)-3)*(1-theta(x(t1)))*(theta(x(t1)));
      Lx2(i,t1)=(alphax(i,j+1)*z(t1,j+1))+(((Axx(i,j)*(1-
          theta(x(t1)))^3)+(Bxx(i,j)*theta(x(t1))*(1-theta(x(
          t1)))^2)+...
        (Cxx(i,j)*(theta(x(t1))^2)*(1-theta(x(t1))))+(Dxx(i,j
          )*(theta(x(t1)))^3))/Qxx);
    else
      Lx(i,t1)=(a(i)*xx(1,t1))+b(i);
      L11(i,t1)=t1+(i-1)*p1;
      Qx=1+(rx(i,j)-3)*(1-theta(xx(t1)))*theta(xx(t1));
      Lx1(i,t1)=(alphax(i,j)*qx1(t1,j))+(((Ax(i,j)*(1-theta(
          xx(t1)))^3)+(Bx(i,j)*theta(xx(t1))*(1-theta(xx(t1))
          )^2)+...
        (Cx(i,j)*(theta(xx(t1))^2)*(1-theta(xx(t1))))+(Dx(i,j
          )*(theta(xx(t1)))^3))/Qx);
      Qxx=1+(rx(i,j)-3)*(1-theta(xx(t1)))*theta(xx(t1));
      Lx2(i,t1)=(alphax(i,j+1)*qx2(t1,j))+(((Axx(i,j)*(1-
          theta(xx(t1)))^3)+(Bxx(i,j)*theta(xx(t1))*(1-theta(
          xx(t1)))^2)+...
        (Cxx(i,j)*(theta(xx(t1))^2)*(1-theta(xx(t1))))+(Dxx(i
          ,j)*(theta(xx(t1)))^3))/Qxx);
    end % end for loop if condition
  end % end for loop t1
  X=[X Lx(i,:)];
  X_na=[X_na L11(i,:)];
  Zx1=[Zx1 Lx1(i,:)]; Zx2=[Zx2 Lx2(i,:)];
end % end for loop i
X1=X;
X_na1=X_na;
Sx1(:,j)=Zx1';
Sx2(:,j)=Zx2';
X=[]; Zx1=[]; Zx2=[];
end % end for loop j
xx=X1;
qx1=Sx1;qx2=Sx2;
Sx1=zeros(length(x)*3^(k+1),3);
Sx2=zeros(length(x)*3^(k+1),3);
p1=length(X1);
qx=[qx1 qx2(:,3)];
```

```
%*****Y-direction*****
%  for  k=1:iter
for  ii =1:3
   for  jj =1:ly -1
      %New generation interpolation points if Fixing f(x(ii), y)
      aa ( jj ) =( y ( jj +1)-y ( jj ) ) /( y ( ly )-y ( 1 ) ) ;
      bb ( jj ) =( ( y ( jj ) *y ( ly ) ) -( y ( jj +1) *y ( 1 ) ) ) /( y ( ly )-y ( 1 ) ) ;
      Ay ( ii , jj ) =z ( ii , jj )-alphay ( ii , jj ) *z ( ii ,1 ) ;
      Dy ( ii , jj ) =z ( ii , jj +1) -( alphay ( ii , jj ) *z ( ii , ly ) ) ;
      By1 ( ii , jj ) =( ry ( ii , jj ) *z ( ii , jj ) ) +hy ( jj ) *dy ( ii , jj ) ;
      By2 ( ii , jj ) =( ry ( ii , jj ) *z ( ii ,1 ) +( y ( ly )-y ( 1 ) ) *dy ( ii ,1 ) ) ;
      By ( ii , jj ) =By1 ( ii , jj )-alphay ( ii , jj ) *By2 ( ii , jj ) ;
      Cy1 ( ii , jj ) =( ry ( ii , jj ) *z ( ii , jj +1) ) -( hy ( jj ) *dy ( ii , jj +1) ) ;
      Cy2 ( ii , jj ) =( ry ( ii , jj ) *z ( ii , ly ) ) -( dy ( ii , ly ) *( y ( ly )-y ( 1 ) ) ) ;
      Cy ( ii , jj ) =Cy1 ( ii , jj )-alphay ( i , j ) *Cy2 ( ii , jj ) ;
      %%New generation interpolation points if Fixing f(x(ii+1),
         y )
      Ayy ( ii , jj ) =( z ( ii +1, jj )-alphay ( ii +1, jj ) *z ( ii +1,1 ) ) ;
      Dyy ( ii , jj ) =( z ( ii +1, jj +1) -( alphay ( ii +1, jj ) *z ( ii +1, ly ) ) ) ;
      Byy1 ( ii , jj ) =( ry ( ii +1, jj ) *z ( ii +1, jj ) ) +hy ( jj ) *dy ( ii +1, jj ) ;
      Byy2 ( ii , jj ) =( ry ( ii +1, jj ) *z ( ii ,1 ) +( y ( ly )-y ( 1 ) ) *dy ( ii ,1 ) ) ;
      Byy ( ii , jj ) =Byy1 ( ii , jj )-alphay ( ii +1, jj ) *Byy2 ( ii , jj ) ;
      Cyy1 ( ii , jj ) =( ry ( ii +1, jj ) *z ( ii +1, jj +1) ) -( hy ( jj ) *dy ( ii +1, jj
         +1) ) ;
      Cyy2 ( ii , jj ) =( ry ( ii +1, jj ) *z ( ii +1, ly ) ) -dy ( ii +1, ly ) *( y ( ly )-y
         ( 1 ) ) ;
      Cyy ( ii , jj ) =Cyy1 ( ii , jj )-alphay ( ii +1, jj ) *Cyy2 ( ii , jj ) ;
   end % End for ' jj '
   y_abvalues =[ aa ' bb ']
   phi=inline ( '( t -0)/8 ') ;
   for  jj =1:ly -1
      for  t2 =1: p2
         if  ( k==1)
            %%New generation interpolation points if Fixing f(x(ii)
               , y )
            Ly ( t2 , jj ) =( aa ( jj ) *y ( t2 ) ) +bb ( jj ) ;
            Lss ( jj , t2 ) =t2 +( jj -1) *p2 ;
            Qy =1+( ry ( ii , jj ) -3) *( 1 - phi ( y ( t2 ) ) ) *phi ( y ( t2 ) ) ;
            Ly1 ( t2 , jj ) =( alphay ( ii , jj ) *z ( ii , t2 ) ) +( ( ( Ay ( ii , jj ) *( 1 - phi
               ( y ( t2 ) ) ) ^3) +( By ( ii , jj ) *phi ( y ( t2 ) ) *( 1 - phi ( y ( t2 ) ) ) ^2)
               +...
               ( Cy ( ii , jj ) *( phi ( y ( t2 ) ) ^2) *( 1 - phi ( y ( t2 ) ) ) ) +( Dy ( ii , jj )
               *( phi ( y ( t2 ) ) ) ^3) ) /Qy ) ;
            %%New generation interpolation points if Fixing f(x(ii
               +1), y )
            Qyy =1+( ry ( ii +1, jj ) -3) *( 1 - phi ( y ( t2 ) ) ) *phi ( y ( t2 ) ) ;
            Ly2 ( t2 , jj ) =( alphay ( ii +1, jj ) *z ( ii +1, t2 ) ) +( ( ( Ayy ( ii , jj )
               *( 1 - phi ( y ( t2 ) ) ) ^3) +( Byy ( ii , jj ) *phi ( y ( t2 ) ) *( 1 - phi ( y (
               t2 ) ) ) ^2) +...
               ( Cyy ( ii , jj ) *( phi ( y ( t2 ) ) ^2) *( 1 - phi ( y ( t2 ) ) ) ) +( Dyy ( ii , jj
               ) *( phi ( y ( t2 ) ) ) ^3) ) /Qyy ) ;
         else
            Ly ( t2 , jj ) =( aa ( jj ) *yy ( 1 , t2 ) ) +bb ( jj ) ;
            Lss ( jj , t2 ) =t2 +( jj -1) *p2 ;
            Qy =1+( ry ( ii , jj ) -3) *( 1 - phi ( yy ( t2 ) ) ) *phi ( yy ( t2 ) ) ;
```

```
Ly1(t2,jj)=(alphay(ii,jj)*qy1(ii,t2))+(((Ay(ii,jj)*(1-
        phi(yy(t2)))^3)+(Byy(ii,jj)*phi(yy(t2))*(1-phi(yy(
        t2)))^2)+...
    (Cy(ii,jj)*(phi(yy(t2))^2)*(1-phi(yy(t2))))+(Dy(ii,jj
        )*(phi(yy(t2)))^3))/Qy);
Qyy=1+(ry(ii+1,jj)-3)*(1-phi(yy(t2)))*phi(yy(t2));
Ly2(t2,jj)=(alphay(ii+1,jj)*qy2(ii,t2))+(((Ayy(ii,jj)
        *(1-phi(yy(t2)))^3)+(Byy(ii,jj)*phi(yy(t2))*(1-phi(
        yy(t2)))^2)+...
    (Cyy(ii,jj)*(phi(yy(t2))^2)*(1-phi(yy(t2))))+(Dyy(ii,
        jj)*(phi(yy(t2)))^3))/Qyy);
    end % End with for if condition k==1'
  end % End with for loop 't2'
  Lyy1=Ly1';
  Lyy2=Ly2';
  Lyy=Ly';
  Y=[Y Lyy(jj,:)];
  Zy1=[Zy1 Lyy1(jj,:)];
  Zy2=[Zy2 Lyy2(jj,:)];
  end % End with for loop 'jj'
 Y1=Y;
 S1(ii,:)=Zy1;
 S2(ii,:)=Zy2;
 Y=[];
 Zy1=[];
 Zy2=[];
end % End with 'ii' (No of iteration)
yy=Y1;
qy1=S1;
qy2=S2;
S1=zeros(3,length(y)*3^(k+1));
S2=zeros(3,length(y)*3^(k+1));
p2=length(Y1);
%end
qy=[qy1 ; qy2(3,:)];
%%*****Surface Evualuation*****
a=zeros(1,N-1);
b=zeros(1,N-1);
c=zeros(1,M-1);
d=zeros(1,M-1);

for  n=1:N-1
  a(n) = (x(n+1)-x(n))/(x(N)-x(1));
  b(n) = ((x(n)*x(N))-(x(n+1)*x(1)))/(x(N)-x(1));
end
for  m=1:M-1
  c(m) = (y(m+1)-y(m))/(y(M)-y(1));
  d(m)= ((y(m)*y(M))-(y(m+1)*y(1)))/(y(M)-y(1));
end
theta=inline('(x-0)/11');
phi=inline('(y-0)/8');
for  ix=1:N-1
  for  i1=1:N1
    if(k==1)
      L1(ix,i1)=a(ix)*x(i1)+b(ix);
```

```
    else
      L1(ix,i1)=a(ix)*XX(i1)+b(ix);
    end % End with if condition
 end % end with for 'i1'
 for jy=1:M-1
   if(ix==1)
     for j1=1:M1
        if(k==1)
          L2(jy,j1)=c(jy)*y(j1)+d(jy);
        else
          L2(jy,j1)=c(jy)*YY(j1)+d(jy);
        end % End with if condition
      end % End with 'j1'
      Y11=[Y11 L2(jy,:)];
    end % End with 'ix==1'
    %% Blending functions
    for i1=1:N1
      for j1=1:M1
        if(k==1)
          gx_1(i1)=((1-theta(x(i1)))^2)*(1+2*theta(x(i1)));%
              (1-theta)^2(1+2*theta)--'po'
          gx_2(i1)=((theta(x(i1)))^2)*(3-2*theta(x(i1)));%theta
              ^3*(3-2*theta)--'p1'
          gy_1(j1)=((1-phi(y(j1)))^2)*(1+2*phi(y(j1)));%(1-phi)
              ^2(1+2*phi)--'qo'
          gy_2(j1)=((phi(y(j1)))^2)*(3-2*phi(y(j1)));%phi
              ^3*(3-2*phi)--'q1'
          R1(i1,j1)=gx_1(i1)*gy_1(j1)*z(ix,jy);
          R2(i1,j1)=gx_1(i1)*gy_2(j1)*z(ix,jy+1);
          R3(i1,j1)=gx_2(i1)*gy_1(j1)*z(ix+1,jy);
          R4(i1,j1)=gx_2(i1)*gy_2(j1)*z(ix+1,jy+1);
          R(i1,j1)=R1(i1,j1)+R2(i1,j1)+R3(i1,j1)+R4(i1,j1);
          L(i1,j1)=(gy_1(j1))*qx(L11(ix,i1),jy)+(gy_2(j1))*qx(
              L11(ix,i1),jy+1)+...
             (gx_1(i1))*qy(ix,Lss(jy,j1))+((gx_2(i1))*qy(ix+1,
                 Lss(jy,j1)))-R(i1,j1);
        else
          gx_1(i1)=((1-theta(XX(i1)))^2)*(1+2*theta(XX(i1)));
          gx_2(i1)=((theta(XX(i1)))^2)*(3-2*theta(XX(i1)));
          gy_1(j1)=((1-phi(YY(j1)))^2)*(1+2*phi(YY(j1)));
          gy_2(j1)=((phi(YY(j1)))^2)*(3-2*phi(YY(j1)));
          R1(i1,j1)=gx_1(i1)*gy_1(j1)*z(ix,jy);
          R2(i1,j1)=gx_1(i1)*gy_2(j1)*z(ix,jy+1);
          R3(i1,j1)=gx_2(i1)*gy_1(j1)*z(ix+1,jy);
          R4(i1,j1)=gx_2(i1)*gy_2(j1)*z(ix+1,jy+1);
          R(i1,j1)=R1(i1,j1)+R2(i1,j1)+R3(i1,j1)+R4(i1,j1);
          L(i1,j1)=(gy_1(j1))*qx(L11(ix,i1),jy)+(gy_2(j1))*qx(
              L11(ix,i1),jy+1)+(gx_1(i1))*qy(ix,Lss(jy,j1))+((
              gx_2(i1))*qy(ix+1,Lss(jy,j1)))-R(i1,j1);
        end
      end
    end
    Z11=[Z11 L];
    L=[];
 end
```

```
    X11 =[ X11  L1 ( ix ,: ) ];
    Z22 =[ Z22 ; Z11 ];
    Z11 =[];
  end
  XX=X11 ; YY=Y11 ; ZZ=Z22 ;
  X11 =[]; Y11 =[]; Z22 =[];
  N1= length (XX) ; M1= length (YY) ;
end
a ; c ;
axis  square ;
surf (XX, YY, ZZ ' ) ;
%title ( ' Surface ' ) ; xlabel ( 'X  values ' ) ;  ylabel ( 'Y  values ' ) ;  zlabel
    ( 'Z  values ' ) ;
hold  off
```

4.1.3 *Numerical Computation*

Let $f_{i,j}^x$ and $f_{i,j}^y$ denote the first partial derivatives of f with respect to x and y respectively. Consider the bivariate Hermite data $\{x_i, \ y_j, \ f_{i,j}, \ f_{i,j}^x, \ f_{i,j}^y : \ i \in \mathbb{N}_M, j \in \mathbb{N}_N\}$ as given in Table 4.1. By choosing the vertical scaling factors and shape parameters (as given in Table 4.2), the graphs of fractal rational cubic FISs are generated and illustrated in Fig. 4.2a. For generating Fig. 4.2a, shape parameters are chosen as $r_x = [1]_{3\times4}$ in x-direction and $r_y = [1]_{4\times3}$ in y-direction. Figure 4.2b represents the bi-cubic partially blended rational FIS for perturbed scaling factors in x-direction and y-direction (given in Table 4.2). Changing the shape parameters r in both directions (x-direction and y-direction), Fig. 4.2c is generated. With the set of scaling factors $\alpha_x = [0]_{3\times4}$ and $\alpha_y = [0]_{4\times3}$, the classical rational cubic surface is developed and it is shown in Fig. 4.2d.

Table 4.1 Hermite interpolation data in the construction of blending rational cubic FISs

$\{(x_i, y_i)\}_{i=1}^4 = \begin{pmatrix} 0 & 0 \\ 2 & 3 \\ 6 & 7 \\ 11 & 8 \end{pmatrix}$	$f = \begin{pmatrix} 1 & 10 & 8 & 11 \\ 2 & 9 & 12 \\ 3 & 12 & 10 & 14 \\ 4 & 13 & 15 \end{pmatrix}$
$f^x = \begin{pmatrix} 4.5 & 1.5 & 0.13 & 1.37 \\ 4.5 & 1.5 & 0.12 & 1.37 \\ 4.5 & 1.5 & 0. & 1.75 \\ 4.45 & 1.61 & 0.25 & 1.25 \end{pmatrix}$	$f^y = \begin{pmatrix} 5 & 2.5 & 0.42 & 0.1 \\ 0.5 & 0.5 & 0.5 & 0.1 \\ 0.5 & 0.5 & 0.5 & 0.5 \\ 0.25 & 0.75 & 0.75 & 0.25 \end{pmatrix}$

Table 4.2 IFSs in the construction of blending rational cubic fractal interpolation surfaces

Scaling parameters	Shape parameters	Figures
$\alpha = \begin{pmatrix} 0.2 & 0.25 & 0.25 & 0.2 \\ 0.35 & 0.3 & 0.35 & 0.34 \\ 0.35 & 0.35 & 0.35 & 0.34 \end{pmatrix}$	$r_x = 0.5*\text{ones}(3,4)$	
$\alpha^* = \begin{pmatrix} 0.124 & 0.562 & 0.30 \\ 0.125 & 0.562 & 0.30 \\ 0.125 & 0.562 & 0.3 \\ 0.125 & 0.562 & 0.3 \end{pmatrix}$	$r_y = 100*\text{ones}(4,3)$	Figure 4.2a
$\alpha = \begin{pmatrix} 0.15 & 0.15 & 0.15 & 0.15 \\ 0.3 & 0.3 & 0.3 & 0.3 \\ 0.45 & 0.45 & 0.45 & 0.45 \end{pmatrix}$	$r_x = 0.5*\text{ones}(3,4)$	
$\alpha^* = \begin{pmatrix} 0.3 & 0.4 & 0.12 \\ 0.3 & 0.4 & 0.12 \\ 0.3 & 0.4 & 0.12 \\ 0.3 & 0.4 & 0.12 \end{pmatrix}$	$r_y = 100*\text{ones}(4,3)$	Figure 4.2b
$\alpha = \begin{pmatrix} 0.2 & 0.25 & 0.25 & 0.2 \\ 0.35 & 0.3 & 0.35 & 0.34 \\ 0.35 & 0.35 & 0.35 & 0.34 \end{pmatrix}$	$r_x = \text{ones}(3,4)$	
$\alpha^* = \begin{pmatrix} 0.124 & 0.562 & 0.30 \\ 0.125 & 0.562 & 0.30 \\ 0.125 & 0.562 & 0.3 \\ 0.125 & 0.562 & 0.3 \end{pmatrix}$	$r_y = \text{ones}(4,3)$	Figure 4.2c
$\alpha = \text{zeros}(3,4)$ $\alpha^* = \text{zeros}(4,3)$	$r_x = 0.5*\text{ones}(3,4)$ $r_y = 100*\text{ones}(4,3)$	Figure 4.2d

4.2 Fractal Surfaces with Variable Scaling

Consider bivariate interpolation data $\{x_i, y_j, f_{i,j}, f_{i,j}^x, f_{x,y}^y : i \in \mathbb{N}_M, j \in \mathbb{N}_N\}$ with increaing x and y values, where $f_{i,j}^x$ and $f_{i,j}^y$ are the x-partial and y-partial derivatives of the original function at (x_i, y_j) respectively. Let $I = [x_1, x_M]$, $J = [y_1, y_N]$, $I_i = [x_i, x_{i+1}]$, $J_j = [y_j, y_{j+1}]$, $h_i = x_{i+1} - x_i$, $h_j = y_{j+1} - y_j$, $D = I \times J$, $D_{i,j} = I_i \times J_j$.

Along the j-th grid line parallel to x-axis
For $j \in \mathbb{N}_N$, $R_j \in \{x_i, f_{i,j}, f_{i,j}^x : i \in \mathbb{N}_M\}$ is the interpolation data along the j-th grid line parallel to x-axis. Consider affine maps $L_i(x) = a_i x + b_i$ defined by $L_i : I \to I_i$ satisfying $L_i(x_1) = x_i$, $L_i(x_M) = x_{i+1}$. The rational fractal interpolation function (RCFIF) [6, 8] is given by

$$\Psi(x, y_j) = \alpha_{i,j}(x)\Psi(L_i^{-1}(x), y_j) + \frac{P_{i,j}(\theta)}{Q_{i,j}(\theta)}, \quad i \in \mathbb{N}_M, \tag{4.3}$$

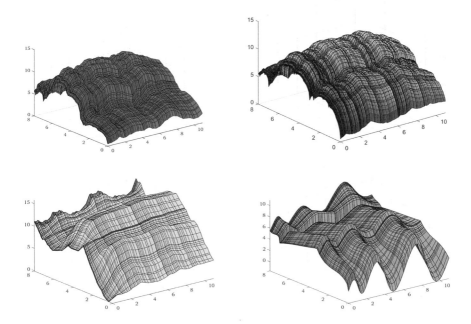

Fig. 4.2 Fractal interpolation surfaces with constant scaling

where $\alpha(x)$ is a Lipschitz function,

$$
\begin{aligned}
P_{i,j}(\theta) = \;& r_{i,j}(f_{i,j} - \alpha_{i,j}(x)f_{1,j})(1 - \theta)^3 + (f_{i+1,j} - \alpha_{i,j}(x)f_{M,j})\theta^3 \\
&+ \{(2r_{i,j} + t_{i,j})f_{i,j} + r_{i,j}h_i f_{i,j}^x - \alpha_{i,j_x}[(2r_{i,j} + t_{i,j})f_{1,j} \\
&+ r_{i,j}(x_M - x_1)f_{1,j}^x]\}\theta(1 - \theta)^2 + \{(r_{i,j} + 2t_{i,j})f_{i+1,j} - t_{i,j}h_i f_{i+1,j}^x \\
&- \alpha_{i,j}(x)[(r_{i,j} + 2t_{i,j})f_{M,j} + r_{i,j}(x_M - x_1)f_{M,j}^x]\}\theta^2(1 - \theta),
\end{aligned}
$$

$$
Q_{i,j}(\theta) = (1 - \theta)r_{i,j} + \theta t_{i,j}, \;\; \theta = \frac{x - x_i}{h_i}, x \in I_i.
$$

Along the i-th grid line parallel to y-axis

For $i \in \mathbb{N}_M$, $R_j \in \{x_i, f_{i,j}, f_{i,j}^x : j \in \mathbb{N}_N\}$ is the interpolation data along the i-th grid line parallel to y-axis. Consider affine maps $L_j^*(y) = c_j y + d_j$ defined by $L_j^* : J \to J_j$ satisfying $L_j^*(y_1) = y_j$, $L_j^*(y_N) = y_{j+1}$. Here $|\alpha_{i,j}^*| < c_j < 1$. We construct RCFIF,

$$
\Psi^*(x_i, y) = \alpha_{i,j}^*(x)\Psi^*(x_i, L_i^{*-1}(y)) + \frac{P_{i,j}^*(\phi)}{Q_{i,j}^*(\phi)}, \;\; j \in \mathbb{N}_N, \tag{4.4}
$$

where $\alpha^*(y)$ is a Lipschitz function,

$$
\begin{aligned}
P_{i,j}^*(\phi) = {} & r_{i,j}(f_{i,j} - \alpha_{i,j}^*(y)f_{i,1})(1-\phi)^3 + (f_{i,j+1} - \alpha_{i,j}^*(y)f_{i,N})\phi^3 \\
& + \{(2r_{i,j} + t_{i,j})f_{i,j} + r_{i,j}h_j f_{i,j}^y - \alpha_{i,j}^*[(2r_{i,j} + t_{i,j})f_{i,1} \\
& + r_{i,j}(y_N - y_1)f_{1,j}^y]\}\phi(1-\phi)^2 + \{(r_{i,j} + 2t_{i,j})f_{i,j+1} - t_{i,j}h_j f_{i,j+1}^y \\
& - \alpha_{i,j}^*(y)[(r_{i,j} + 2t_{i,j})f_{i,N} + r_{i,j}(y_N - y_1)f_{i,N}^y]\}\phi^2(1-\phi), \\
Q_{i,j}(\phi) = {} & (1-\phi)r_{i,j} + \phi t_{i,j}, \quad \phi = \frac{y - y_j}{h_j}, \ y \in J_j.
\end{aligned}
$$

The construction of univariate FIFs $\Psi(x, y_j)$, $\Psi(x, y_{j+1})$ and $\Psi^*(x_i, y)$, $\Psi^*(x_{i+1}, y)$ are discussed with cubic Hermite functions in the following. Let

$$
\begin{aligned}
b_{0,3}^i(x) &= (1-\theta)^2(1+2\theta), \\
b_{3,3}^i(x) &= \theta^2(3-2\theta), \\
b_{0,3}^j(y) &= (1-\phi)^2(1+2\phi), \\
b_{3,3}^j(y) &= \phi^2(3-2\phi).
\end{aligned}
$$

On each individual patch $D_{i,j} = I_i \times J_j$, $i \in N_{m-1}$, $j \in N_{n-1}$ (see Fig. 4.1), a blending rational cubic spline FIS is defined using the blending coons technique as

$$
\Phi(x, y) = -\begin{bmatrix} -1 & b_{0,3}^i(x) & b_{3,3}^i(x) \end{bmatrix} \begin{bmatrix} 0 & \Psi(x, y_j) & \Psi(x, y_{j+1}) \\ \Psi^*(x_i, y) & f_{i,j} & f_{i,j+1} \\ \Psi^*(x_{i+1}, y) & f_{i+1,j} & f_{i+1,j+1} \end{bmatrix} \begin{bmatrix} -1 \\ b_{0,3}^j(y) \\ b_{3,3}^j(y) \end{bmatrix}.
$$

The function interpolation surface Φ interpolate the given data at grid points. The MATLAB code of rational cubic FIS with variable scalings is illustrated in the following.

```
%Fractal  Surface  q(X)=P(X)/Q(X),  where  P(X)  is  cubic  and  Q(X)  is
    linear.
%Here  scaling  factor  is  alpha  and  two  shape  parameters  arw  mu,nu.
%Inputs    are  (x,y,z,alphax,alphay,mu,nu,partial  derivatives(dx,
    dy)).
%Outputs  will  be  "fractal  surface".
clc;clear  all;
close  all;format('short')
%*****Given  Data******
x=[0  4  8  10];y=[0  3  5  9];
iter=3;
grid  on
m=length(x);
n=length(y);
%z=[0.4  9  5  10;1  10  6  11;2  11  7  12;3  12  8  13];
```

```
z =[3  11  9  8  ;4  8  10  7;1  10  12  4;4  12  14  17];
%z=z ';
%z=[4  9  100  110  :  2  7   20  210  :  150  300  30  1135; 285  315  350  1410
     ];
lx=length(x);
ly=length(y);
N=length(x);
M=length(y);
p1=lx ;
p2=ly ;
dx=[4.5  1.5  0.125  1.375;4.5  1.5  0.125  1.375;4.5  1.5  0.25
     1.75;4.3929  1.6071  0.25  1.25];
%dx=dx '
dy=[0.5  0.5  0.4167  0.0833;0.5  0.5  0.4167  0.0833;0.5  0.5  0.5
     0.5;0.25  0.75  0.75  −0.25];
%dy=dy '
%%
% Free  Shape  parameters%%%%%%%%%%%%%%%
r_x =[1  1  1  1;  1  1  1  1;  1  1  1  1];mux=r_x ;
t_x =[1  1  1  1;1  1  1  1;1  1  1  1];nux=t_x ;
%r_x =10*[1  1  1  1;  1  1  1  1;  1  1  1  1];mux=r_x ;
%r_x =100*[1  1  1  1;  1  1  1  1;  1  1  1  1];mux=r_x ;
%t_x =[100  100  100  100;1  1  1  1;100  100  100  100];nux=t_x ;
 r_y =[1  1  1;1  1  1;1  1  1;1  1  1];muy=r_y ;
 %  r_y =100*[1  1  1;1  1  1;1  1  1;  1  1  1];muy=r_y ;
 %r_y =[1  8   4;1  1  1;100  1  1;1  1  1];muy=r_y ;
 t_y =[1  1  1;  1  1  1;  1  1  1;  1  1  1];nuy=t_y ;
%t_y =[100  1  1;  100  1  1;  1  1  1;  1  1  1];nuy=t_y ;
for  i =1:lx −1
     h(i)=x(i+1)−x(i);
end
for  l =1:ly −1
     hy(l)=y(l+1)−y(l);
end
%****** Definitions  of  Matrices ******
a=zeros (1 , lx −1);
b=zeros (1 , lx −1);
Lx =[];
Lx1 =[]; Lx2 =[];
X_na =[];
X1 =[]; Zx1 =[];
X =[];
Zx2 =[];
a=zeros (1 , ly −1);
b=zeros (1 , ly −1);
Lyy =[];
Ly1 =[]; Ly2 =[];
Y1 =[];
Y =[];
Zy1 =[]; Zy2 =[];
L =[]; L1 =[]; L2 =[];
L22 =[]; L222 =[];
X11 =[]; Y11 =[];
Z11 =[]; Z22 =[];
XX =[]; YY =[]; ZZ =[];
```

```
N1=N;
M1=M;
%%%%%%%%%%%%%%%%%%%%%%%%%%%%%%%%%%%%%%
iter=2;
%%%%%%%%%%%%%%%%%%% (km)Finding coffients
for k=1:iter
    %*******X-direction(Y is fixed)******

    for j=1:3
        for i=1:lx-1
            %%-----Finding coffients if Fixing f(x,y(j))
            a(i)=(x(i+1)-x(i))/(x(lx)-x(1))
            b(i)=((x(i)*x(lx))-(x(i+1)*x(1)))/(x(lx)-x(1));
            %%%
            %-----Variable Alpha-----
            %% Finding alpha values Here all  rows are same,
                because all rows are approximatively same.
            if(k==1)
                for t1=1:p1
                    if i==1
                        alphax(i,t1)=x(t1)/(x(p1)-x(1))*(1/260)
                        %alphax(i,t1)=x(t1)/(3*[x(p1)-x(1)]);
                    elseif i==2
                        alphax(i,t1)=sin(x(t1)/(270*x(p1)-x(1)))
                        %alphax(i,t1)=2*cos(x(t1))/(x(p1)-x(1));
                    elseif i==3
                        alphax(i,t1)=abs(log(1+x(t1)/x(p1)-x(1)))
                            *(1/250)
                        %alphax(i,t1)=abs(log(1+x(t1))/x(p1)-x
                            (1))
                    end
                end
            else
                if i==1
                    alphax(i,t1)=X1(t1)/(X1(p1)-X1(1))*(1/260);
                    %alphax(i,t1)=X1(t1)/(3*[X1(p1)-X1(1)]);
                elseif i==2
                    alphax(i,t1)=sin(X1(t1)/(270*(X1(p1)-X1(1))))
                        ;
                    %alphax(i,t1)=2*cos(X1(t1))/(X1(p1)-X1(1));
                else
                    alphax(i,t1)=abs(log(1+X1(t1)/(X1(p1)-X1(1)))
                        )*(1/250);
                    %alphax(i,t1)=abs(log(1+X1(t1))/(X1(p1)-X1(1)
                        ))
                end
            end
            %-------------------------------------------------
            Ax(i,j)=(z(i,j)-alphax(i,j)*z(1,j))*mux(i,j);
            Dx(i,j)=(z(i+1,j)-(alphax(i,j)*z(lx,j)))*nux(i,j);
            %x valuxe
            Bx1(i,j)=((2*mux(i,j)+nux(i,j))*z(i,j))+(mux(i,j)*h(i
                )*dx(i,j));
            Bx2(i,j)=(2*mux(i,j)+nux(i,j))*z(1,j)+(mux(i,j)*(x(lx
                )-x(1))*dx(1,j));
```

```
Bx(i,j)=Bx1(i,j)-alphax(i,j)*Bx2(i,j);
% Dx Value
Cx1(i,j)=((2*nux(i,j)+mux(i,j))*z(i+1,j))-(nux(i,j)*h
    (i)*dx(i+1,j));
Cx2(i,j)=((2*nux(i,j)+mux(i,j))*z(lx,j))-(nux(i,j)*dx
    (lx,j)*(x(lx)-x(1)));
Cx(i,j)=Cx1(i,j)-alphax(i,j)*Cx2(i,j);
% END DX valuxe
%====End finding  coffients if Fixing f(x,y(j))
%-----Finding coffients if Fixing f(x, y(j+l))
Axx(i,j)=(z(i,j+1)-(alphax(i,j+1)*z(1,j+1)))*mux(i,j
    +1);
Dxx(i,j)=(z(i+1,j+1)-(alphax(i,j+1)*z(lx,j+1)))*nux(i
    ,j+1);
%Cx value
Bxx1(i,j)=((2*mux(i,j+1)+nux(i,j+1))*z(i,j+1))+(mux(i
    ,j+1)*h(i)*dx(i,j+1));
Bxx2(i,j)=(2*mux(i,j+1)+nux(i,j+1))*z(1,j+1)+(mux(i,j
    +1)*(x(lx)-x(1))*dx(1,j+1));
Bxx(i,j)=Bxx1(i,j)-alphax(i,j+1)*Bxx2(i,j);
% Dx value
Cxx1(i,j)=((2*nux(i,j+1)+mux(i,j+1))*z(i+1,j+1))-(nux
    (i,j+1)*h(i)*dx(i+1,j+1));
Cxx2(i,j)=((2*nux(i,j+1)+mux(i,j+1))*z(lx,j+1))-(nux(
    i,j+1)*dx(lx,j+1)*(x(lx)-x(1)));
Cxx(i,j)=Cxx1(i,j)-alphax(i,j+1)*Cxx2(i,j);
% END DX value
%%%===End finding  coffients if Fixing f(x,y(j+l))
end% end for loop i
 theta=inline('(x-0)/11');
%theta=inline('(x-0)/6');
for i=1:lx-1
    for tl=1:pl
        if (k==1)
            Lx(i,tl)=(a(1,i)*x(1,tl))+b(1,i);
            L11(i,tl)=tl+(i-1)*pl;
            %Finding coffients if Fixing f(x, y(j))
            % alphax value ------------------
            Qx1=mux(i,j)*(1-theta(x(tl)));
            Qx2=nux(i,j)*(theta(x(tl)));
            Lx1(i,tl)=(alphax(i,tl)*z(tl,j))+((Ax(i,j)
                *(1-theta(x(tl)))^3)+(Bx(i,j)*theta(x(tl)
                )*(1-theta(x(tl)))^2)+...
                (Cx(i,j)*(theta(x(tl))^2)*(1-theta(x(tl))
                ))+(Dx(i,j)*(theta(x(tl)))^3))/(Qx1+
                Qx2);
            %Finding coffients if Fixing f(x, y(j+l))
            Qxx1=mux(i,j+1)*(1-theta(x(tl)));
            Qxx2=nux(i,j+1)*(theta(x(tl)));
             Lx2(i,tl)=(alphax(i,tl)*z(tl,j+1))+((Axx(i,j
                )*(1-theta(x(tl)))^3)+(Bxx(i,j)*theta(x(
                tl))*(1-theta(x(tl)))^2)+...
                (Cxx(i,j)*(theta(x(tl))^2)*(1-theta(x(tl)
                )))+(Dxx(i,j)*(theta(x(tl)))^3))/(
                Qxx1+Qxx2);
```

```
            else
                Lx(i,t1)=(a(i)*xx(1,t1))+b(i);
                L11(i,t1)=t1+(i-1)*p1;
                Qx1=mux(i,j)*(1-theta(xx(t1)));
                Qx2=nux(i,j)*(theta(xx(t1)));

                %alphax(i,t1)=0;
                %%% alpha values
                if i==1
                    alphax(i,t1)=X1(t1)/(X1(p1)-X1(1))
                        *(1/260);
                elseif i==2
                    alphax(i,t1)=sin(X1(t1)/(270*(X1(p1)-X1
                        (1))));
                else
                    alphax(i,t1)=abs(log(1+X1(t1)/(X1(p1)-X1
                        (1))))*(1/250);
                end
                %%%%%%%%%%
                Lx1(i,t1)=(alphax(i,t1)*qx1(t1,j))+(((Ax(i,j)
                    *(1-theta(xx(t1)))^3)+(Bx(i,j)*theta(xx(
                    t1))*(1-theta(xx(t1)))^2)+...
                    (Cx(i,j)*(theta(xx(t1))^2)*(1-theta(xx(t1
                        ))))+(Dx(i,j)*(theta(xx(t1)))^3))/(
                        Qx1+Qx2));
                Qxx1=mux(i,j+1)*(1-theta(xx(t1)));
                Qxx2=nux(i,j+1)*(theta(xx(t1)));
                Lx2(i,t1)=(alphax(i,t1)*qx2(t1,j))+(((Axx(i,j
                    )*(1-theta(xx(t1)))^3)+(Bxx(i,j)*theta(xx
                    (t1))*(1-theta(xx(t1)))^2)+...
                    (Cxx(i,j)*(theta(xx(t1))^2)*(1-theta(xx(
                        t1))))+(Dxx(i,j)*(theta(xx(t1)))^3))
                        /(Qxx1+Qxx2));
            end % end for loop if condition
        end % end for loop t1
        X=[X Lx(i,:)];
        X_na=[X_na L11(i,:)];
        Zx1=[Zx1 Lx1(i,:)];
        Zx2=[Zx2 Lx2(i,:)];
    end % end for loop i
    X1=X;
    X_na1=X_na;
    Sx1(:,j)=Zx1';
    Sx2(:,j)=Zx2';
    X=[];
    Zx1=[];
    Zx2=[];
end % end for loop k(iteration)

xx=X1;
qx1=Sx1;
qx2=Sx2;
Sx1=zeros(length(x)*3^(k+1),3);
Sx2=zeros(length(x)*3^(k+1),3);
```

```
p1=length(X1);
qx=[qx1  qx2(:,3)];
% End for fix 'y' and change 'x'.
%******Y-direction******
%for k=1:iter
for  ii =1:3
    for  jj =1:ly -1
        % Finding coffients if Fixing f(x(ii), y)
        aa( jj )=(y( jj +1)-y( jj ))/( y( ly )-y(1));
        bb( jj )=((y( jj )*y( ly ))-(y( jj +1)*y(1)))/( y( ly )-y(1));

        %%
        if (k==1)
            for  t2 =1:p2
                if  jj ==1
                    alphay( t2 , jj )=sin (y( jj )/( y( p2 )-y(1)))
                        +0.01;
                elseif jj ==2
                    alphay( t2 , jj )=abs ( sec (y( jj )-y(1)))
                        *(1/120);
                else
                    alphay( t2 , jj )=exp (y( jj )/120)*(1/103);
                end
            end
        else
            for  t2 =1:p2
                if  jj ==1
                    alphay( t2 , jj )=sin (Y1( jj )/( Y1( p2 )-Y1(1)))
                        +0.01;
                elseif jj ==2
                    alphay( t2 , jj )=abs ( sec (Y1( jj )-Y1(1)))
                        *(1/120);
                else
                    alphay( t2 , jj )=exp (Y1( jj )/120)*(1/103);
                end
            end
        end
%
        Ay( ii , jj )=(z( ii , jj )-alphay( ii , jj )*z( ii ,1))*muy( ii , jj )
            ;
        Dy( ii , jj )=(z( ii , jj +1)-(alphay( ii , jj )*z( ii , ly )))*nuy(
            ii , jj );
%----
        By1( ii , jj )=((2*muy( ii , jj )+nuy( ii , jj ))*z( ii , jj ))+(muy(
            ii , jj )*hy( jj )*dy( ii , jj ));
        By2( ii , jj )=(2*muy( ii , jj )+nuy( ii , jj ))*z( ii ,1)+(muy( ii ,
            jj )*(y( ly )-y(1))*dy( ii ,1));
        By( ii , jj )=By1( ii , jj )-alphay( ii , jj )*By2( ii , jj );
%----
        Cy1( ii , jj )=((2*nuy( ii , jj )+muy( ii , jj ))*z( ii , jj +1))-(
            nuy( ii , jj )*hy( jj )*dy( ii , jj +1));
        Cy2( ii , jj )=((2*nuy( ii , jj )+muy( ii , jj ))*z( ii , ly ))-(nuy(
            ii , jj )*dy( ii , ly )*(y( ly )-y(1)));
        Cy( ii , jj )=Cy1( ii , jj )-alphay( ii , jj )*Cy2( ii , jj );
%-----Finding coffients if Fixing f(x(ii+1), y)
```

```
            Ayy(ii,jj)=(z(ii+1,jj)-alphay(ii+1,jj)*z(ii+1,1))*muy
                (ii+1,jj);
            Dyy(ii,jj)=(z(ii+1,jj+1)-(alphay(ii+1,jj)*z(ii+1,ly))
                )*nuy(ii+1,jj);
            Byy1(ii,jj)=((2*muy(ii+1,jj)+nuy(ii+1,jj))*z(ii+1,jj)
                )+(muy(ii+1,jj)*hy(jj)*dy(ii+1,jj));
            Byy2(ii,jj)=(2*muy(ii+1,jj)+nuy(ii+1,jj))*z(ii,1)+(
                muy(ii,jj)*(y(ly)-y(1))*dy(ii,1));
            Byy(ii,jj)=Byy1(ii,jj)-alphay(ii+1,jj)*Byy2(ii,jj);
            %------
            Cyy1(ii,jj)=((2*nuy(ii+1,jj)+muy(ii+1,jj))*z(ii+1,jj
                +1))-(nuy(ii+1,jj)*hy(jj)*dy(ii+1,jj+1));
            Cyy2(ii,jj)=((2*nuy(ii+1,jj)+muy(ii+1,jj))*z(ii+1,ly)
                )-(nuy(ii+1,jj)*dy(ii+1,ly)*(y(ly)-y(1)));
            Cyy(ii,jj)=Cyy1(ii,jj)-alphay(ii+1,jj)*Cyy2(ii,jj);
            %checking complete                 %------
    end % End for 'jj'
        phi=inline('(t-0)/8');
    %phi=inline('(t-0)/8');
    for jj=1:ly-1
        for t2=1:p2
            if (k==1)
                %Finding coffients if Fixing f(x(ii), y)
                Ly(t2,jj)=(aa(jj)*y(t2))+bb(jj);
                Lss(jj,t2)=t2+(jj-1)*p2;
                %=====
                Qy1=muy(ii,jj)*(1-phi(y(t2)));
                Qy2=nuy(ii,jj)*(phi(y(t2)));
                Ly1(t2,jj)=(alphay(t2,jj)*z(ii,t2))+((Ay(ii,
                    jj)*(1-phi(y(t2)))^3)+(By(ii,jj)*phi(y(t2
                    ))*(1-phi(y(t2)))^2)+...
                    (Cy(ii,jj)*(phi(y(t2))^2)*(1-phi(y(t2))))
                        +(Dy(ii,jj)*(phi(y(t2)))^3))/(Qy1+Qy2
                        );

                %Finding coffients if Fixing f(x(ii+1), y)
                Qyy1=muy(ii+1,jj)*(1-phi(y(t2)));
                Qyy2=nuy(ii+1,jj)*(phi(y(t2)));
                Ly2(t2,jj)=(alphay(t2,jj)*z(ii+1,t2))+((Ayy(
                    ii,jj)*(1-phi(y(t2)))^3)+(Byy(ii,jj)*phi(
                    y(t2))*(1-phi(y(t2)))^2)+...
                    (Cyy(ii,jj)*(phi(y(t2))^2)*(1-phi(y(t2)))
                        )+(Dyy(ii,jj)*(phi(y(t2)))^3))/(Qyy1+
                        Qyy2);
                %--------------------

            else

                Ly(t2,jj)=(aa(jj)*yy(1,t2))+bb(jj);
                Lss(jj,t2)=t2+(jj-1)*p2;
                Qy1=muy(ii,jj)*(1-phi(yy(t2)));
                Qy2=nuy(ii,jj)*(phi(yy(t2)));
                %%------------------------ alpha is zero
                % alphay(t2,jj)=0
                %%%%%%%%%%%%%%%%%%%%%%%%%%%%%%%%%%%%%%%%%%%%
```

```
                         Ly1(t2,jj)=(alphay(t2,jj)*qy1(ii,t2))+(((Ay(
                               ii,jj)*(1-phi(yy(t2)))^3)+(Byy(ii,jj)*phi
                               (yy(t2))*(1-phi(yy(t2)))^2)+...
                               (Cy(ii,jj)*(phi(yy(t2))^2)*(1-phi(yy(t2))
                                    ))+(Dy(ii,jj)*(phi(yy(t2)))^3))/(Qy1+
                                    Qy2));
                         Qyy1=muy(ii+1,jj)*(1-phi(yy(t2)));
                         Qyy2=nuy(ii+1,jj)*(phi(yy(t2)));
                         Ly2(t2,jj)=(alphay(t2,jj)*qy2(ii,t2))+(((Ayy(
                               ii,jj)*(1-phi(yy(t2)))^3)+(Byy(ii,jj)*phi
                               (yy(t2))*(1-phi(yy(t2)))^2)+...
                               (Cyy(ii,jj)*(phi(yy(t2))^2)*(1-phi(yy(t2)
                                    )))+(Dyy(ii,jj)*(phi(yy(t2)))^3))/(
                                    Qyy1+Qyy2));
                 end % End with for if condition k==1'
             end % End with for loop 't2'
             Lyy1=Ly1';
             Lyy2=Ly2';
             Lyy=Ly';
             Y=[Y Lyy(jj,:)];
             Zy1=[Zy1 Lyy1(jj,:)];
             Zy2=[Zy2 Lyy2(jj,:)];
         end % End with for loop 'jj'
         Y1=Y;
         S1(ii,:)=Zy1;
         S2(ii,:)=Zy2;
         Y=[];
         Zy1=[];
         Zy2=[];
end % End with 'ii' (No of iteration)
yy=Y1;
qy1=S1;
qy2=S2;
S1=zeros(3,length(y)*3^(k+1));
S2=zeros(3,length(y)*3^(k+1));
p2=length(Y1);
%end
qy=[qy1 ; qy2(3,:)];

%*******Surface Evualuation******
a=zeros(1,N-1);
b=zeros(1,N-1);
c=zeros(1,M-1);
d=zeros(1,M-1);

for  n=1:N-1
     a(n) = (x(n+1)-x(n))/(x(N)-x(1));
     b(n) = ((x(n)*x(N))-(x(n+1)*x(1)))/(x(N)-x(1));
end
for  m=1:M-1
     c(m) = (y(m+1)-y(m))/(y(M)-y(1));
     d(m)= ((y(m)*y(M))-(y(m+1)*y(1)))/(y(M)-y(1));
end
theta=inline('(x-0)/11');
phi=inline('(y-0)/8');
```

```
for  ix =1:N−1
    for  i1 =1:N1
        if ( k ==1)
            L1( ix , i1 )=a( ix )*x( i1 )+b( ix );
        else
            L1( ix , i1 )=a( ix )*XX( i1 )+b( ix );
        end % End with if condition
    end % end with for 'i1'
    for  jy =1:M−1
        if ( ix ==1)
            for  j1 =1:M1
                if ( k ==1)
                    L2( jy , j1 )=c( jy )*y( j1 )+d( jy );
                else
                    L2( jy , j1 )=c( jy )*YY( j1 )+d( jy );
                end % End with if condition
            end % End with 'j1'
            Y11 =[ Y11  L2( jy ,:) ];
        end % End with 'ix ==1'
        for  i1 =1:N1
            for  j1 =1:M1
                if ( k ==1)
                    gx_1( i1 )=((1 − theta ( x( i1 ))) ^2)*(1+2* theta (
                        x( i1 )));% (1− theta )^2(1+2* theta )−−'po

                    gx_2( i1 )=(( theta ( x( i1 ))) ^2)*(3 −2* theta ( x(
                        i1 )));%theta ^3*(3 −2* theta )−−'p1'
                    gy_1( j1 )=((1 − phi ( y( j1 ))) ^2)*(1+2* phi ( y( j1
                        )));%(1− phi )^2(1+2* phi )−−'qo'
                    gy_2( j1 )=(( phi ( y( j1 ))) ^2)*(3 −2* phi ( y( j1 ))
                        );%phi ^3*(3 −2* phi )−−'q1'
                    R1( i1 , j1 )=gx_1( i1 )*gy_1( j1 )*z( ix , jy );
                    R2( i1 , j1 )=gx_1( i1 )*gy_2( j1 )*z( ix , jy +1);
                    R3( i1 , j1 )=gx_2( i1 )*gy_1( j1 )*z( ix +1, jy );
                    R4( i1 , j1 )=gx_2( i1 )*gy_2( j1 )*z( ix +1, jy +1);
                    R( i1 , j1 )=R1( i1 , j1 )+R2( i1 , j1 )+R3( i1 , j1 )+R4
                        ( i1 , j1 );
                    L( i1 , j1 )=(gy_1( j1 ))*qx( L11( ix , i1 ), jy )+(
                        gy_2( j1 ))*qx( L11( ix , i1 ), jy +1) +...
                        ( gx_1( i1 ))*qy( ix , Lss ( jy , j1 )) +(( gx_2(
                            i1 ))*qy( ix +1,Lss ( jy , j1 )))−R( i1 , j1
                            );
                else
                    gx_1( i1 )=((1 − theta (XX( i1 ))) ^2)*(1+2* theta
                        (XX( i1 )));
                    gx_2( i1 )=(( theta (XX( i1 ))) ^2)*(3 −2* theta (
                        XX( i1 )));
                    gy_1( j1 )=((1 − phi (YY( j1 ))) ^2)*(1+2* phi (YY(
                        j1 )));
                    gy_2( j1 )=(( phi (YY( j1 ))) ^2)*(3 −2* phi (YY( j1
                        )));
                    R1( i1 , j1 )=gx_1( i1 )*gy_1( j1 )*z( ix , jy );
                    R2( i1 , j1 )=gx_1( i1 )*gy_2( j1 )*z( ix , jy +1);
                    R3( i1 , j1 )=gx_2( i1 )*gy_1( j1 )*z( ix +1, jy );
                    R4( i1 , j1 )=gx_2( i1 )*gy_2( j1 )*z( ix +1, jy +1);
```

```
                              R(i1 ,j1)=R1(i1 ,j1)+R2(i1 ,j1)+R3(i1 ,j1)+R4
                                 (i1 ,j1);
                              L(i1 ,j1)=(gy_1(j1))*qx(L11(ix ,i1),jy)+(
                                 gy_2(j1))*qx(L11(ix ,i1),jy+1)+(gx_1(
                                 i1))*qy(ix ,Lss(jy ,j1))+((gx_2(i1))*qy
                                 (ix+1,Lss(jy ,j1)))−R(i1 ,j1);
                      end
                  end
              end
              Z11=[Z11  L];
              L=[];
          end
          X11=[X11  L1(ix ,:)];
          Z22=[Z22;Z11];
          Z11=[];
      end
      XX=X11;
      YY=Y11;
      ZZ=Z22;
      X11=[];
      Y11=[];
      Z22=[];
      N1=length(XX);
      M1=length(YY);
end
axis  square;
surf(XX,YY,ZZ');
%title('Surface');
%xlabel('X values');ylabel('Y values');zlabel('Z values');
hold  off
```

4.2.1   Numerical Computation

Let $f_{i,j}^x$ and $f_{i,j}^y$ denote the first partial derivatives of f with respect to x and y respectively. Consider the bi-variate Hermite data $\{x_i, \ y_j, \ f_{i,j}, \ f_{i,j}^x, \ f_{i,j}^y : i \in \mathbb{N}_M, j \in \mathbb{N}_N\}$ as given in Table 4.3. By choosing the vertical scaling functions (see Table 4.5) and shape parameters (as given in Table 4.4), the graphs of fractal rational cubic FISs are generated and illustrated in Fig. 4.3. For generating Fig. 4.3a, shape parameters are chosen as $r_x = [1]_{3 \times 4}$ in x-direction and $r_y = [1]_{4 \times 3}$ in y-direction. Figure 4.3b represents the bi-cubic partially blended rational FIS for perturbed shape parameters in x-direction (given in Table 4.4). Figure 4.3c represents the rational FIS by changing shape parameters in y-direction (given in Table 4.4). Changing both vertical scaling vectors and shape parameters, Fig. 4.3d is generated. Figure 4.3e represents the rational FIS by changing α and t_x (as in Table 4.4). Changing the vertical scaling vectors (α) and r_y, t_x, Fig. 4.3f is generated.

Table 4.3 Hermite interpolation data in the construction of blending rational cubic FISs

$$\{(x_i, y_i)\}_{i=1}^4 = \begin{pmatrix} 0 & 0 \\ 4 & 3 \\ 8 & 5 \\ 10 & 9 \end{pmatrix} \qquad f = \begin{pmatrix} 3 & 11 & 9 & 8 \\ 4 & 8 & 10 & 7 \\ 1 & 10 & 12 & 4 \\ 4 & 12 & 14 & 17 \end{pmatrix}$$

$$f^x = \begin{pmatrix} 4.5 & 1.5 & 0.125 & 1.375 \\ 4.5 & 1.5 & 0.125 & 1.375 \\ 4.5 & & & \\ 1.5 & & 0.25 & 1.75 \\ 4.3929 & 1.607 & 10.25 & 1.25 \end{pmatrix} \qquad f^y = \begin{pmatrix} 0.5 & 0.5 & 0.4167 & 0.0833 \\ 0.5 & 0.5 & 0.4167 & 0.0833 \\ 0.5 & 0.5 & 0.5 & 0.5 \\ 0.25 & 0.75 & 0.75 & -0.25 \end{pmatrix}$$

Table 4.4 Shape parameters in the construction of RCFIFs

Shape parameters	Figures
$r_x = $ ones(3,4)	Figure 4.3a, c, e, f
$t_x = $ ones(3,4)	Figure 4.3a, c, d
$r_y = $ ones(4,3)	Figure 4.3a–e
$t_y = $ ones(4,3)	Figure 4.3a–f
$r_x = 10*$ones(3,4)	Figure 4.3b
$r_x = 100*$ones(3,4)	Figure 4.3d
$r_y = 100*$ones(4,3)	Figure 4.3f
$t_x = \begin{pmatrix} 100 & 100 & 100 & 100 \\ 1 & 1 & 1 & 1 \\ 100 & 100 & 100 & 100 \end{pmatrix}$	Figure 4.3b, e, f
$r_y = \begin{pmatrix} 1 & 8 & 4 \\ 1 & 1 & 1 \\ 100 & 1 & 1 \\ 1 & 1 & 1 \end{pmatrix}$	Figure 4.3c
$t_y = \begin{pmatrix} 100 & 1 & 1 \\ 100 & 1 & 1 \\ 1 & 1 & 1 \\ 1 & 1 & 1 \end{pmatrix}$	Figure 4.3c

Table 4.5 Scaling factors in the construction of RCFIFs

Scaling factors	Figures
$\alpha = \left[\frac{x}{260(x_n - x_1)}, \frac{\sin(x)}{270*(x_n - x_1)}, \frac{|\log(1+x)|}{250(x_n - x_1)} \right]$	Figure 4.3a–c
$\alpha^* = \left[\frac{\sin(y)}{y(n) - y(1)} + 0.01, \frac{|\sec(y) - y(1)|}{120}, \frac{e^y}{120} * \frac{1}{103} \right]$	Figure 4.3a–f
$\alpha = \left[\frac{x}{3*(x_n - x_1)}, \frac{2\cos(x)}{x_n - x_1}, \frac{|\log(1+x(t1)|}{x_n - x_1} \right]$	Figure 4.3d–f

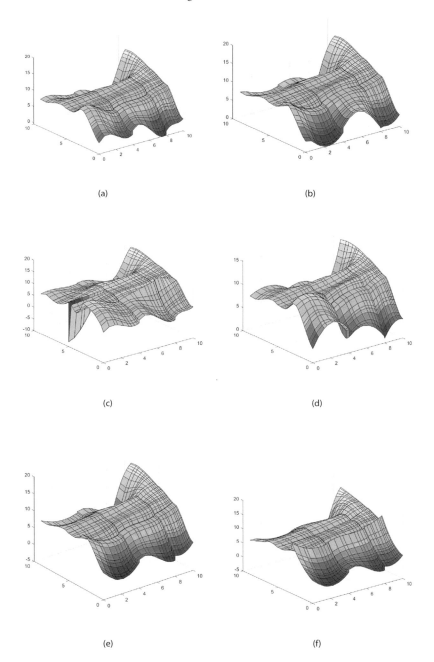

Fig. 4.3   Fractal interpolation surfaces with variable scaling

References

1. K.M. Reddy, G. Saravana Kumar, A.K.B. Chand, Family of shape preserving fractal-like be'zier curves. Fractals **28**(06), 2050105 (2022)
2. H. Xie, H. Sun, The study on bivariate fractal interpolation functions and creation of fractal interpolated surfaces. Fractals **5**(04), 625–634 (1997)
3. P. Bouboulis, L. Dalla, Fractal interpolation surfaces derived from fractal interpolation functions. J. Math. Anal. Appl. **336**(2), 919–936 (2007)
4. M.A. Navascués, R.N. Mohapatra, M.N. Akhtar, Construction of fractal surfaces. Fractals **28**(02), 2050033 (2020)
5. M.G.P. Prasad, M.N. Akhtar, Fractal interpolation surfaces and perturbations on vertical scaling factors. Int. J. Nonlinear Sci. **21**(1), 3–12 (2016)
6. K.M. Reddy, A.K.B. Chand, Constrained univariate and bivariate rational fractal interpolation. Int. J. Comput. Methods Eng. Sci. Mech. **20**(5), 404–422 (2019)
7. A.K.B. Chand, P. Viswanathan, K.M. Reddy, A novel approach to surface interpolation: marriage of coons technique and univariate fractal functions. Math. Anal. Appl. **143**, 577–592 (2015)
8. A.K.B. Chand, K.M. Reddy, Constrained fractal interpolation functions with variable scaling. Sib. Elektron. Mat. Izv **15**, 60–73 (2018)

Chapter 5
Applications

In this section, applications of fractal interpolation function are discussed, in particular, patterns of mountains and clouds are approximated and the positive cases of Omicron are reconstructed.

5.1 Patterns of Mountains and Clouds

Consider the following three sets of interpolation data

- $\{(0, 0), (2, 2), (4, 0), (6.5, 0.5), (10, 0.1)\}$,
- $\{(3, 1.75), (3.5, 2.05), (4, 2.25), (5, 2.35), (5.45, 2.15), (5.85, 1.95), (5.5, 1.6),$ $(4.5, 1.53), (3.5, 1.45), (3, 1.75)\}$,
- $\{(6.75, 2), (7.2.25), (7.75, 2.5), (8.75, 2.6), (9.15, 2.4), (9.75, 2.2),$ $(9.25, 1.75), (8.25, 1.65), (7, 1.8), (6.75, 2)\}$.

An example problem is considered to demonstrate the effect of scaling factors in visualizing mountains and clouds. Figure 5.1 illustrates geometric models of mountains and clouds. The graphical data points marked are to be interpolated to give the picture of mountains and clouds. The conventional linear interpolation is used with varying scalings to obtain Fig. 5.1a–d. The varying scalings are provided in Table 5.1. A desirable effect is obtained by trial and error using scaling factor values for the data points generating the cloud and mountains, for more details refer [1].

© The Author(s), under exclusive license to Springer Nature Switzerland AG 2023
S. Banerjee et al., *Fractal Patterns with MATLAB*,
SpringerBriefs in Complexity,
https://doi.org/10.1007/978-3-031-48102-4_5

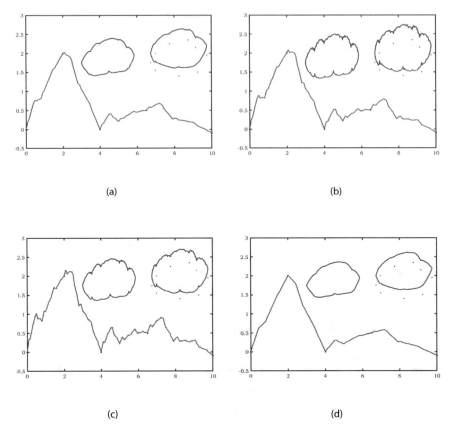

(a)   (b)

(c)   (d)

Fig. 5.1 Geometric models of mountains and clouds

Table 5.1 Scaling parameters associated with mountains and clouds

Figures	Scalings α_1	Scalings α_2
Figure 5.1a	0.15	0.1
Figure 5.1b	0.2	0.15
Figure 5.1c	0.25	0.3
Figure 5.1d	0.1	0.25

5.1.1 MATLAB Simulation

```
%% Constructing  the  geometric  models  using  three  affine  maps
function  []= Geometric_model ()
clc ; clear  all ; close  all ;
format  'short'
%%Data1
x1 =[0  2  4  6.5  10]; y1 =[0  2  0  0.5  −0.1]
```

```
lx = length(x1);
%alpha1 =0.15*ones(1,lx −1);%figure
%alpha1 =0.2*ones(1,lx −1);%figure1
%alpha1 =0.25*ones(1,lx −1);%figure2
alpha1 =0.1*ones(1,lx −1);%figure3
iter =6;
[X1 Y1]= Affine_FIF(x1,y1,alpha1,iter);
plot (x1,y1,'.k','markersize',10);hold on
plot(X1,Y1,'r−');hold on
%%Data2
xx =[3 3.5 4 5 5.45 5.85 5.5 4.5 3.5 3];
yy =[1.75 2.05 2.25 2.35 2.15 1.95 1.6 1.53 1.45 1.75]
xx1 =[3 3.5 4 5 5.45 5.85];yy1 =[1.75 2.05 2.25 2.35 2.15 1.95]
xx2 =[3 3.5 4.5 5.5 5.85];yy2 =[1.75 1.45 1.53 1.6 1.95];
lx1 = length(xx1)
%alpha2 =0.15*ones(1,lx1 −1);%figure
%alpha2 =0.3*ones(1,lx1 −1);%figure1
%alpha2 =0.25*ones(1,lx1 −1);%figure2
alpha2 =0.1*ones(1,lx1 −1);%figure3
[XX1 YY1]= Affine_FIF(xx1,yy1,alpha2,iter);
[XX2 YY2]= Affine_FIF(xx2,yy2,alpha2,iter);
plot (xx,yy,'.k');hold on
plot(XX1,YY1,'b−');hold on
plot(XX2,YY2,'b−')
%%Data3
xxx1 =[6.75 7 7.75 8.75 9.15 9.75];
yyy1 =[1.75 2 2.25 2.35 2.15 1.95 1.5 1.4 1.55 1.75]
xxx1 =[6.75 7 7.75 8.75 9.15 9.75];yyy1 =[2 2.25 2.5 2.6 2.4
    2.2];
xxx2 =[6.75 7 8.25 9.25 9.75];yyy2 =[2 1.8 1.65 1.75 2.2];
[XXX1 YYY1]= Affine_FIF(xxx1,yyy1,alpha2,iter);
[XXX2 YYY2]= Affine_FIF(xxx2,yyy2,alpha2,iter);
plot (xxx,yyy,'.k');hold on
plot(XXX1,YYY1,'b−');hold on
plot(XXX2,YYY2,'b−')
end
function [X1 Y1]= Affine_FIF(x,y,alpha,iter)
lx =length(x);
N=lx ;
for i=1:lx −1
   diff_x(i)=x(i+1)−x(i);length_x =x(N)−x(1);
   a(i)=diff_x(i)/length_x ;
   b(i)=(x(N)*x(i)−x(1)*x(i+1))/length_x ;
   c(i)=[y(i+1)−y(i)−alpha(i)*(y(N)−y(1))]/length_x ;
   d(i)=[x(N)*y(i)−x(1)*y(i+1)−alpha(i)*(x(N)*y(1)−x(1)*y(N))]/
      length_x ;
end
abcd_values =[a' b' c' d']
%%%%%%%%%%%%%%%%%%%%%%%%%%%%%%%%%
L=[];L1 =[];X1 =[];Y1 =[];X=[];Y=[];
p=N;
for k=1:iter
   for i =1:N−1
      for t1 =1:p
         if (k==1)% First iteration
```

```
% Input data is (x,y) or given data
L(i,t1)=a(i)*x(t1)+b(i);
L1(i,t1)=alpha(i)*y(t1)+c(i)*x(t1)+d(i);
else % More than one iteration
% Input data is (X1,Y1) or output after the first
iteration
L(i,t1)=a(i)*X1(t1)+b(i);
L1(i,t1)=alpha(i)*Y1(t1)+c(i)*X1(t1)+d(i);
end
end
X=[X L(i,:)];
Y=[Y L1(i,:)];
end
X1=X;      Y1=Y;      X=[];      Y=[];      g=[X1' Y1'];
g=str2num(num2str(g,10));g=unique(g,'rows');
X1=g(:,1);Y1=g(:,2);      [X1 Y1]; p=length(X1);
end
end
```

This illustration shows that the effect of scaling factor on the shape of the inter-polation is not very apparent and the designer has to exercise several iterations for fine tuning the parameters to obtain the desired effects.

5.2 Reconstruction of Omicron Data

The seven-days moving average of daily positive cases of Omicron for the five countries, namely India, Italy, South Africa, UK and USA are considered. The affine fractal interpolation function discussed in Chap. 2 is used to reconstruct the graphs of seven-days moving average of Omicron cases. Table 5.2 provides the time duration of the data associated with first and second waves of Omicron. The data points $\{(x_i, y_i)\}$ represent the time (x-axis) and the value of seven-days moving average (y-axis). The data corresponding to the fractal graphs of Omicron in Figs. 5.2 and 5.3 are taken from [2]. Interested readers many consult [3], where the fractal graphs of Omicron are analysed and using a moving average model the successive waves of COVID-19 are predicted.

Table 5.2 Duration of first and second waves of Omicron for five countries

Country	First wave	Second wave
India	01.04.2020–31.01.2021	01.03.2021–31.07.2021
South Africa	01.05.2020–31.10.2020	01.12.2020–31.03.20211
USA	01.09.2020–28.02.2021	01.08.2021–31.03.2021
UK	01.10.2020–31.03.2021	01.12.2021–30.04.2021
Italy	01.10.2020–31.05.2021	01.11.2021–31.05.2022

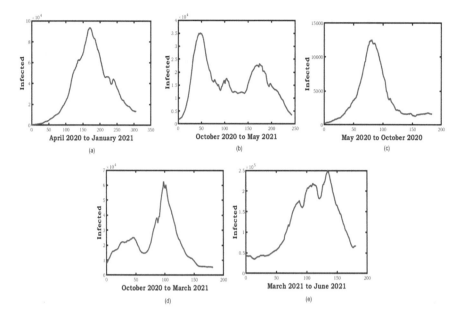

Fig. 5.2 Fractal transformation of COVID data for first wave: **a** India, **b** South Africa, **c** USA, **d** UK and **e** Italy

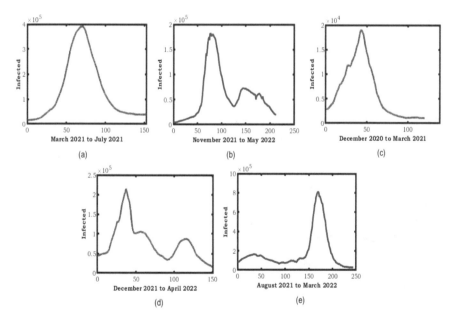

Fig. 5.3 Fractal transformation of COVID data for second wave: **a** India, **b** South Africa, **c** USA, **d** UK and **e** Italy

5.2.1   MATLAB Simulation

The MATLAB code for the reconstruction of Omicron data using the affine fractal intepolation function is provided, here the scalings are chosen such that $|\alpha_i| < 1$.

```
%% Affine Fractal Interpolation Function
%% L_i (x)=a_i x+b_i
%% F_i(x,y)=alpha_i (x) *y +Q_i(x),
% where Q_i(x) is the affine function
function []=Const_Affine_FIF()
clc;clear all;close all;
format 'short'
%x=[0   1/3  1/2  2/3  1];y=[1   3  5/2  3.5  3/2];% Data
filename = 'India_covid_wave1.xlsx';
Data=xlsread(filename);
%size(Data)
x=Data(:,1);y=Data(:,2);
x1=Data(1:153,3);
y1=Data(1:153,4);
%iter=input('Enter the number of iterations:=');
iter=1;
lx=length(x);
ly=length(y);
lx1=length(x1)
ly1=length(y1)
[X1 Y1]=Affine_FIF(x,y,iter);
[X2 Y2]=Affine_FIF(x1,y1,iter);
figure
subplot(1,2,1)
plot(X1,Y1,'r-');
xlabel('April2020 to January 2021');
ylabel('Infected')
subplot(1,2,2)
plot(X2,Y2,'r-');
xlabel('March2021 to July 2021');
ylabel('Infected')
end
function [X1 Y1]=Affine_FIF(x,y,iter)
lx=length(x);
alpha=0.0033*ones(1,lx-1);
N=lx;
for i=1:lx-1
   diff_x(i)=x(i+1)-x(i);length_x=x(N)-x(1);
   a(i)=diff_x(i)/length_x;
   b(i)=(x(N)*x(i)-x(1)*x(i+1))/length_x;
   c(i)=[y(i+1)-y(i)-alpha(i)*(y(N)-y(1))]/length_x;
   d(i)=[x(N)*y(i)-x(1)*y(i+1)-alpha(i)*(x(N)*y(1)-x(1)*y(N))]/
        length_x;
end
abcd_values=[a' b' c' d'];
%%%%%%%%%%%%%%%%%%%%%%%%%%%%%%%%
L=[];L1=[];X1=[];Y1=[];X=[];Y=[];
p=N;
for k=1:iter
```

```
for  i =1:N−1
  for  t1 =1:p
    if  (k==1)% First iteration
      L( i , t1 )=a ( i ) *x ( t1 )+b ( i ) ;
      L1 ( i , t1 )=alpha ( i ) *y ( t1 )+c ( i ) *x ( t1 )+d ( i ) ;
    else % More than one iteration
      % Input data is (X1,Y1) or output after the first
        iteration
      L( i , t1 )=a ( i ) *X1 ( t1 )+b ( i ) ;
      L1 ( i , t1 )=alpha ( i ) *Y1 ( t1 )+c ( i ) *X1 ( t1 )+d ( i ) ;
    end
  end
  X=[X  L( i ,:) ];
  Y=[Y  L1 ( i ,:) ];
end
X1=X;      Y1=Y;      X =[];      Y =[];      g =[X1 ’  Y1 ’];
g=str2num ( num2str ( g ,10 ) ) ; g=unique ( g , ´rows ´) ;
X1=g ( : ,1 ) ; Y1=g ( : ,2 ) ;
[X1  Y1 ];  p=length (X1) ;
end
end
% %%%%%%%%%%%%%%%%%%%%%%%%%%%%%%%%%%%%%%
```

References

1. K.M. Reddy, N. Vijender, A fractal model for constrained curve and surface. Eur. Phys. J.: Spec. Top. **232**, 1015–102518 (2023)
2. E. Mathieu, H. Ritchie, L. Rod's-Guirao, et al., Coronavirus pandemic (COVID-19). Our World in Data (2020). https://ourworldindata.org/covid-cases
3. A. Gowrisankar, T.M.C. Priyanka, S. Banerjee, Omicron: a mysterious variant of concern. Eur. Phys. J. Plus **137**(1), 1–8 (2022)

Printed in the United States
by Baker & Taylor Publisher Services